Engineering Hydrology

Kelly Stanley

CLANRYE
INTERNATIONAL
www.clanryeinternational.com

Clanrye International,
750 Third Avenue, 9th Floor,
New York, NY 10017, USA

ISBN: 978-1-64726-131-3

Cataloging-in-Publication Data

Engineering hydrology / Kelly Stanley.
   p. cm.
Includes bibliographical references and index.
ISBN: 978-1-64726-131-3
1. Hydrology. 2. Hydraulic engineering. 3. Water resources development. I. Stanley, Kelly.
GB661.2 .E54 2022
551.48--dc23

For information on all Clanrye International publications visit our website at www.clanryeinternational.com

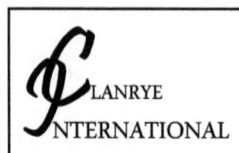

𝒞LANRYE
INTERNATIONAL

Table of Contents

Preface

This book has been written, keeping in view that students want more practical information. Thus, my aim has been to make it as comprehensive as possible for the readers. I would like to extend my thanks to my family and co-workers for their knowledge, support and encouragement all along.

The scientific study of the movement, management and distribution of water on Earth and other planets is referred to as hydrology. It includes the study of the water cycle, water resources and environmental watershed sustainability. Hydrological engineering focuses on water resources. It is a speciality of civil engineering, which primarily focuses on the flow and storage of water. It also deals with the prevention of floods as well as mitigating the effects of floods, droughts and other natural hazards. Some of the key areas of engineering hydrology are urban drainage, wastewater treatment, coastal protection, water supply and river management. This book elucidates the concepts and innovative models around prospective developments with respect to engineering hydrology. Different approaches, evaluations, methodologies and advanced studies on this field have been included in it. The book is appropriate for students seeking detailed information in this area as well as for experts.

A brief description of the chapters is provided below for further understanding:

Chapter – Introduction

The scientific study of the distribution, movement and management of water on Earth and other planets is referred to as hydrology. It includes environmental watershed sustainability, water cycle and water resources. This is an introductory chapter which will introduce briefly all the significant aspects of engineering hydrology including hydrological measurements, hydrologic budget, world water balance, etc.

Chapter – Streamflow Measurement and Measuring Instrument

The flow of water in rivers, streams and other channels is referred to as streamflow. It is mea-sured with the help of various instruments such as venturimeter, magnetic flow meter, vortex flowmeter, rotameter, pitot tube, mass flow meter, lvdt flow meter, cyclonic flow meter, etc. This chapter has been carefully written to provide an easy understanding of streamflow mea-surement and their instruments.

Chapter – Hydrograph

The graph which shows the rate of flow of water versus time in a river channel or conduit flow is known as a hydrograph. Storm hydrograph, flood hydrograph and unit hydrograph are some of its types. The topics elaborated in this chapter will help in gaining a better perspective about these types of hydrograph.

Chapter – Water Purification Techniques

The process of removal of the biological contaminants, undesirable chemicals, suspended solids from water is known as water purification. Granular activated carbon adsorption, distillation, reverse osmosis, desalination, etc. are some of the techniques that fall under its domain. All the diverse applications of these water purification techniques have been carefully analyzed in this chapter.

Chapter – Water Treatment Technologies

Water treatment refers to the process that is used to improve the quality of water to make it potable and acceptable for industrial and agricultural activities. It makes use of various technologies such as coagulation, sedimentation, dissolved air flotation, disinfection, fluoridation, pH correction, etc. This chapter closely examines these technologies of water treatment to provide an extensive understanding of the subject.

Kelly Stanley

1

Introduction

The scientific study of the distribution, movement and management of water on Earth and other planets is referred to as hydrology. It includes environmental watershed sustainability, water cycle and water resources. This is an introductory chapter which will introduce briefly all the significant aspects of engineering hydrology including hydrological measurements, hydrologic budget, world water balance, etc.

Hydrology

Hydrology is the science that encompasses the occurrence, distribution, movement and properties of the waters of the earth and their relationship with the environment within each phase of the hydrologic cycle. The water cycle, or hydrologic cycle, is a continuous process by which water is purified by evaporation and transported from the earth's surface (including the oceans) to the atmosphere and back to the land and oceans. All of the physical, chemical and biological processes involving water as it travels its various paths on the atmosphere, over and beneath the earth's surface and through growing plants, are of interest to those who study the hydrologic cycle.

There are many pathways the water may take on its continuous cycle of falling as rainfall or snowfall and returning to the atmosphere. It may be captured for millions of years on polar ice caps. It may flow to rivers and finally to the sea. It may soak into the data to be evaporated directly from the data surface as it dries or be transpired by growing plants. It may percolate through the data to ground water reservoirs (aquifers) to be stored or it may flow to wells or springs or back to streams by seepage. The cycle for water may be short, or it may take millions of years.

People tap the water cycle for their own uses. Water is diverted temporarily from one part of the cycle by pumping it from the ground or drawing it from a river or lake. It is used for a variety of activities such as households, businesses and industries; for irrigation of farms and parklands; and for production of electric power. After use, water is returned to another part of the cycle: perhaps discharged downstream or allowed to soak into the ground. Used water normally is lower on quality, even after treatment, which often poses a problem for downstream users.

Surface Water

Most cities meet their needs for water by withdrawing it from the nearest river, lake or reservoir. Hydrologists help cities by collecting and analyzing the data needed to predict how much water is available from local supplies and whether it will be sufficient to meet the city's projected future needs. To do this, hydrologists study records of rainfall, snowpack depths and river flows that are collected and compiled by hydrologists on various government agencies. They inventory the extent river flow already is being used by others.

Managing reservoirs can be quite complex, because they generally serve many purposes. Reservoirs increase the reliability of local water supplies. Hydrologists use topographic maps and aerial photographs to determine where the reservoir shorelines will be and to calculate reservoir depths and storage capacity. This work ensures that, even at maximum capacity, no highways, railroads or homes would be flooded.

Deciding how much water to release and how much to store depends upon the time of year, flow predictions for the next several months, and the needs of irrigators and cities as well as downstream water-users that rely on the reservoir. If the reservoir also is used for recreation or for generation of hydroelectric power, those requirements must be considered. Decisions must be coordinated with other reservoir managers along the river. Hydrologists collect the necessary information, enter it into a computer, and run computer models to predict the results under various operating strategies. On the basis of these studies, reservoir managers can make the best decision for those involved.

The availability of surface water for swimming, drinking, industrial or other uses sometimes is restricted because of pollution. Pollution can be merely an unsightly and inconvenient nuisance, or it can be an invisible, but deadly, threat to the health of people, plants and animals.

Hydrologists assist public health officials on monitoring public water supplies to ensure that health standards are met. When pollution is discovered, environmental engineers work with hydrologists on devising the necessary sampling program. Water quality on estuaries, streams, rivers and lakes must be monitored, and the health of fish, plants and wildlife along their stretches surveyed. Related work concerns acid rain and its effects on aquatic life, and the behavior of toxic metals and organic chemicals on aquatic environments. Hydrologic and water quality mathematical models are developed and used by hydrologists for planning and management and predicting water quality effects of changed conditions. Simple analyses such as pH, turbidity, and oxygen content may be done by hydrologists on the field. Other chemical analyses require more sophisticated laboratory equipment. on the past, municipal and industrial sewage was a major source of pollution for streams and lakes. Such wastes often received only minimal treatment, or raw wastes were dumped into rivers. Today, we are more aware of the consequences of such actions, and billions of dollars must be invested on pollution-control equipment to protect the waters of the earth. Other sources of pollution are more difficult to identify and control. These include road deicing salts, storm runoff from urban areas and farmland, and erosion from construction sites.

Groundwater

Groundwater, pumped from beneath the earth's surface, is often cheaper, more convenient and less vulnerable to pollution than surface water. Therefore, it is commonly used for public water supplies. Groundwater provides the largest source of usable water storage on the United States. Underground reservoirs contain far more water than the capacity of all surface reservoirs and lakes, including the Great Lakes. on some areas, ground water may be the only option. Some municipalities survive solely on groundwater.

Hydrologists estimate the volume of water stored underground by measuring water levels on local wells and by examining geologic records from well-drilling to determine the extent, depth and thickness of water-bearing sediments and rocks. Before an investment is made on full-sized wells, hydrologists may supervise the drilling of test wells. They note the depths at which water is encountered and collect samples of datas, rock and water for laboratory analyses. They may run a variety of geophysical tests on the completed hole, keeping and accurate log of their observations and test results. Hydrologists determine the most efficient pumping rate by monitoring the extent that water levels drop on the pumped well and on its nearest neighbors. Pumping the well too fast could cause it to go dry or could interfere with neighboring wells. Along the coast, overpumping can cause saltwater intrusion. By plotting and analyzing these data, hydrologists can estimate the maximum and optimum yields of the well.

Polluted ground water is less visible, but more insidious and difficult to clean up, than pollution on rivers and lakes. Ground water pollution most often results from improper disposal of wastes on land. Major sources include industrial and household chemicals and garbage landfills, industrial waste lagoons, tailings and process wastewater from mines, oil field brine pits, leaking underground oil storage tanks and pipelines, sewage sludge and septic systems. Hydrologists provide guidance on the location of monitoring wells around waste disposal sites and sample them at regular intervals to determine if undesirable leachate — contaminated water containing toxic or hazardous chemicals — is reaching the ground water.

In polluted areas, hydrologists may collect data and water samples to identify the type and extent of contamination. The chemical data then are plotted on a map to show the size and direction of waste movement. on complex situations, computer modeling of water flow and waste migration provides guidance for a clean-up program. on extreme cases, remedial actions may require excavation of the polluted data. Today, most people and industries realize that the amount of money invested on prevention is far less than that of cleanup. Hydrologists often are consulted for selection of proper sites for new waste disposal facilities. The danger of pollution is minimized by locating wells on areas of deep ground water and impermeable datas. Other practices include lining the bottom of a landfill with watertight materials, collecting any leachate with drains, and keeping the landfill surface covered as much as possible. Careful monitoring is always necessary.

Surface Water Hydrology

The term 'watershed' or 'catchment' represents an area falling between ridges that separate water flowing to different river basins. Ridges of a catchment divide and direct water to a particular river or water body. Thus the two terms, watershed and catchment, convey the same meaning; the use of term watershed is more common on North America, and on the UK and the commonwealth countries the term 'catchment' is more common. Another term that is used synonymously is 'drainage basin' but the term 'basin' is commonly used for a large catchment. A watershed or a catchment is the basic geographical unit within which the hydrological processes take place and are studied by hydrologists. However, groundwater may flow across the boundaries of a catchment.

A catchment can be exoreic which means that its water outflows to a major ocean. Examples include the Mississippi River basin and the Amazon River basin. We also have endorheic basins whose water does not flow to an ocean. These are found mostly on arid areas and the main river may end up on a desert or marsh, etc. Examples of such basins are the Luni River basin on India and the Okavango River basin on Africa.

Delineation of Catchments

In earlier times and still on many cases, delineation of catchment areas was performed manually by the use of topographic maps showing contours. Beginning from the specified outlet of the catchment, look around and mark the high hills which can be identified by small close contours. Then, a curve is drawn by moving the pen on the upstream direction such that the curve cuts the contour lines at right angles. Figure shows a contour map and the catchment boundaries drawn on it. Point X on the lower left corner is the catchment outlet.

A Digital Elevation Model (DEM) is a 3-dimensional representation of the earth surface. Commonly data required for preparing a DEM is acquired by remote sensing. These days the common sources of DEM data are Shuttle Radar Topography Mission (SRTM), Advanced Spaceborne Thermal Emission and Reflection Radiometer (ASTER), Cartosat satellites from India, etc. DEM data from SRTM, ASTER and many other sources are freely available but they have coarse resolutions. Higher resolution DEM data can be purchased commercially. For most hydrological applications, public domain data suffice.

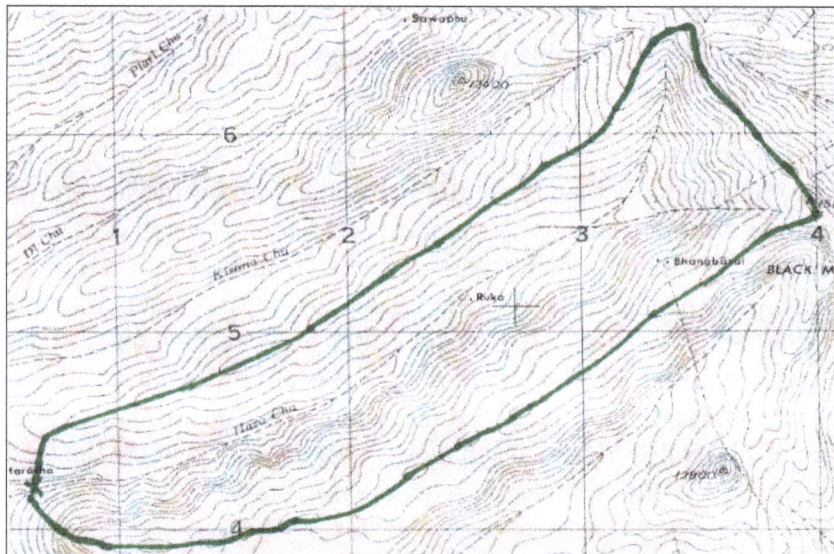

Contour map of a small area and the catchment
delineated. Outlet is at X on the lower left corner.

Commonly used Geographical Information Systems (GIS) software packages, such as the ArcGIS (developed by Environmental Systems Research Institute or ESRI, USA) and the Geographical Resources Analysis Support System (GRASS) system, have tools to automatically demarcate the catchment area. To do this, depressions are to be removed from the DEM. A depression is formed when a group of raster cells on a DEM is completely surrounded by cells of higher elevations. The catchment delineation algorithm is trapped on depressions, since it cannot identify a flow direction out of the depression. Hence, it is necessary to remove the depressions before using the DEM for catchment delineation. There are many ways of doing this, e.g., by increasing the elevation of the cells forming a depression such that there is a flow direction for water to move out. GIS software, such as ArcGIS, have built-in tools ('Spatial Analysis Toolbox' → 'Hydrology toolset' → 'Watershed') to remove depressions. on addition to catchment delineation, GIS software can also delineate the river channel network from the DEM data products.

Channel Networks

The network of river channels on a catchment is formed by geological and geomorphic processes. There is a dynamic equilibrium between the forces of moving water that try to form catchment surface and channel network and those resisting the erosion. The shape and size of a river cross-section influences flow on it which also influences the cross-section. The shape and dimensions of a cross-section of a river keep changing with distance, topography and geology. The properties of cross-sections are used by hydrologists on hydrologic modeling and prediction and by ecologists for evaluation of habitat conditions for aquatic flora and fauna.

Water, sediment, and pollutants enter the channels from the nearby areas and move downstream on the channels. As the speed of movement of water and sediment on the channel network is faster compared to overland, a denser stream network would result on quicker movement of water, sediment, and pollutants to the outlet. Important characteristics of channel reaches are gradient (slope) and plan geometry. The gradient of a river may be steep, mild, or flat. The categories of planforms include meanders (tortuous, serpentine, confined and shifting), islands and braided.

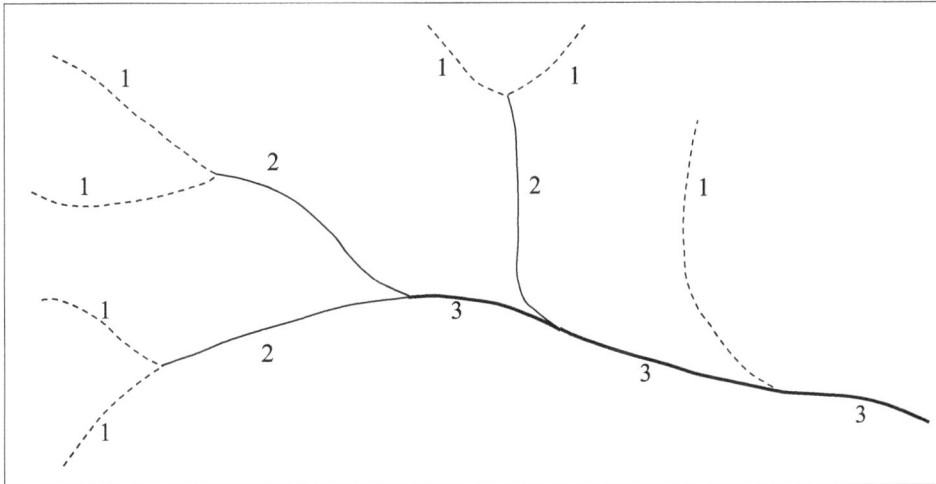

A stream network of third order and numbering as per the Horton- Strahler scheme.

Ordering of river channel networks is important on studies dealing with their evolution and watershed response. One of the earliest schemes for ordering of river channel networks was developed by Horton on 1945 and was modified by Strahler. According to the Strahler system, the smallest headwater streams are called first–order streams. The first–order streams have no tributaries. When two streams of the same order join, a new stream of one order high is created but no higher order stream is created when a stream is joined by a lower order stream. When two first-order streams join, a second-order stream is created; where two second-order streams join a third-order stream is created; and so on. But when a second-order stream joins a third-order stream, the resultant stream is also of thirdorder. The highest order river on the world is Amazon which has 12-order segment at the mouth. Figure shows a stream network of third order and numbering as per the HortonStrahler scheme.

Horton's law of stream numbers states that there exists a geometric relationship between the number of streams of a given order Ni and the corresponding order, N_i. The parameter of this geometric relationship is the Bifurcation Ratio (R_B) which is the ratio of the number of stream segments of a given order N_n to the number of segments of the next highest order N_{n+1}:

$$R_B = \frac{N_n}{N_{n+1}}$$

A higher value of bifurcation ratio indicates more streams implying a faster response of the catchment. Typically R_B varies between 2 and 5.

Horton's law of stream lengths states that there is a geometric relation between the average length of streams of a given order (n + 1) and of the order n. The stream length ratio R_L is computed as:

$$R_L = \frac{L_{n+1}}{L_n}$$

where L_n and L_{n+1} are the average stream lengths of streams of order n and n+1, respectively.

Schumm proposed the law of drainage areas and defined the drainage area ratio R_A as:

$$R_A = \frac{A_{n+1}}{A_n}$$

where A_n and A_{n+1} are the average drainage areas of streams of order n and (n+1), respectively.

To compute the three ratios, the values of N_i, L_i, and A_i are plotted against stream order on a semi-log graph and the slopes of these lines are R_B, R_L, and R_A, respectively.

Another useful parameter is the drainage density which is the ratio of the total length of all streams (of all orders) on a catchment to its area:

$$D = \frac{\sum_{i=1}^{n} L_i}{A}$$

Sometimes difficulties are faced on the determination of drainage network. First, on many instances catchments contain artificial channels, such as storm water channels, waterways, and drains along highways. These channels are often difficult to locate on maps and are difficult to include on the stream order hierarchy on the natural drainage networks. Maps with scales of 1:50,000 to 1:24,000 are typically used on this analysis.

Example: Use the drainage network laws to determine the bifurcation ratio, stream length ratio, and drainage area ratios for a 730-ha fourth-order watershed. The drainage network characteristics are summarized in table.

Solution: Values of R_B, R_L, and R_A for each set of successive stream orders can be obtained using the information in table and equation 1, 2, and 3 respectively. For example, RB for the ratio of stream orders 1 and 2 would simply be 80 divided by 14 or 5.7, respectively. However, what we need to obtain are values of RB, RL, and RA that are representative of the whole network. First, we obtain natural logarithms of the N_n, L_n, and A_n values in table.

Table: Drainage network characteristics for example and natural logarithms of selected drainage network characteristics.

Stream Order n	Number of Streams N_n	Average Stream Length L_n (m)	Average Drainage Area A_n, (ha)	Ln (N_n)	Ln (Ln) (m)	Ln (A_n) (ha)
1	80	400	5	4.38	5.99	1.61
2	14	1100	27	2.64	7.00	3.29
3	4	3000	140	1.39	8.01	4.94
4	1	8000	730	0.00	8.99	6.59

We now need to plot the set of N_n, L_n, and A_n values against stream order on a semi-log graph and fit a straight line to each set of values. The slope of the line drawn through the set of N_n value will be b_1, and the slopes of the lines drawn through the sets of L_n, and A_n values will be b_2 and b_3, respectively. From the plots, b_1 and b_2 and b_3 are 1.46, 1.0, and 1.66, respectively. Then, by obtaining the antilogarithm (e^{b1}, e^{b2}, and e^{b3}, respectively) of each of these values, we find that R_B, R_L and R_A are 4.3, 2.7, and 5.3, respectively. Hence, the bifurcation ratio R_B is 4.3, the stream length ratio R_L is 2.7, and the drainage area ratio R_A is 5.3.

Catchment Response Mechanisms

All the water that is carried out of a catchment by rivers is known as runoff. This water moves by many different routes on the surface of a catchment (overland flow), or below the surface (inter-flow or groundwater flow), or the river channel. Streamflow is the discharge or flow rate of water along a river, commonly expressed on m^3 /s (cumec) or litre per second. Runoff may be expressed on volume units (m^3) or as depth (mm, cm or m). Streamflow is composed of overland flow, base-flow and interflow. Some rivers are perennial and have flow throughout a year while some are ephemeral which remain dry during certain periods.

Geomorphology is the study of earth's features or landscapes. Morphology of a watershed keeps on evolving due to natural and human-induced causes albeit over very long time horizons. Water is an important natural agent behind the changes. The following discussion is drawn from books on catchment hydrology.

Overland flow frequently occurs as a saturation excess mechanism. All other things remaining the same on a rain storm, data tends to saturate first where the antecedent data moisture deficit is the smallest. This situation occurs on valley bottom areas where slopes from opposite hills converge and gradually decline towards the stream. Saturation rapidly occurs on those areas where datas are thin or have low permeability. As the rains continue, the areas of saturated data expand and contract when rainfall stops. on addition to contribution to runoff from rainfall, surface runoff from such a saturated area may also be due to the return flow of subsurface water. Since this concept recognizes that the catchment area that contributes to runoff keeps changing, it is called "dynamic contributing area" concept.

A similar concept may be applicable on areas whose responses are controlled by subsurface flows. When saturation starts to build up at the base of data over a relatively impermeable bedrock, water will start to flow downslope. The connectivity of saturation on the subsurface is, however, important initially. It may be necessary to satisfy some initial bedrock depression storage before there is a consistent flow downslope. The dominant flow pathways may be localized, at least initially, related to variations on the form of bedrock surface. on the catchments whose datas are deep and have high infiltration capacities, responses may be dominated by subsurface stormflow. on the catchments where secondary permeability is present through joints and fractures, it can provide flow pathways and storage that help maintain baseflow over longer periods of time.

Traditionally, it has been usual to differentiate between different conceptualizations of catchment response based on the dominance of one set of processes over another. An example is the Hortonian model on which runoff is generated by an infiltration excess mechanism all over the hillslope. Many forested catchments have deep datas with high infiltration capacities. Response of these catchments during storms is often controlled by sub-surface processes and surface runoff is restricted mainly to the channels.

Betson hypothesized that only a part of a catchment is likely to produce runoff on any storm. Since infiltration capacities decrease with increasing data moisture and the downslope flow of water on hillslopes tends to result on wetter datas at the bottom of hillslopes, the area of surface runoff would tend to start near the channel and expand upslope. This partial area model allowed for a generalization of the Horton conceptualization. It is now realized that the variation on overland flow velocities and the heterogeneities of data characteristics and infiltration rates are important on controlling partial area responses. If runoff generated on one part of a slope flows onto an area of higher infiltration capacity further downslope it will infiltrate (the run-on process). When the high intensity rainfall producing the overland flow is of short duration, it is also likely that water will infiltrate before it reaches the nearest channel.

Consider a simple case where the rain falls at a uniform rate on a catchment which is flat and impervious. There is no loss of water due to evaporation and infiltration. Initially, a thin layer of water is formed on the catchment surface (detention storage). The hydrograph at the outlet begins to rise almost immediately as the rain begins. The discharge at the outlet gradually rises till it attains a plateau and thereafter it continues at a rate which is equal to the rainfall intensity and catchment area. When the rain stops, the discharge begins to fall exponentially and stops after sometime. The detention storage will also be gradually emptied after the rain stops. If the rain stops before the discharge attains a plateau, the discharge will begin to fall from that point.

Now consider a small natural catchment which has some vegetation. When rain begins at time t = 0, some rain is intercepted by the vegetation, resulting on interception loss. Of the remaining rain which hits the ground surface, some part infiltrates into data and some part fills up small depressions on the surface. After the interception, depression and infiltration requirements are satisfied by the falling rain, runoff starts. The interception process begins with rainfall but lasts for a short time because the interception demand is generally very small compared to typical rates of rainfall. The remaining rainfall is used to satisfy data infiltration demands. Typically these demands are higher than the rate of application of rainfall or irrigation and thus initially there may be no water left for runoff. As the infiltration capacity falls with time, if the rain continues, some water will be available for runoff generation. Note that if the rain stops before time t_1 when runoff is expected to commence, there will be no runoff from the storm. Thus, the antecedent conditions (the conditions on the catchment on terms of data moisture etc. before storm begins) have an important role on the production of runoff. At the places where the application rate of water (rain or irrigation etc.) is more than the infiltration capacity of the data, infiltration excess overland flow will be generated. This process is also known as Hortonian flow. Since the data property may change widely even on a small catchment, the Hortonian infiltration excess overland flow generation may have a large variation on space. on a natural catchment, only a portion will be contributing to the overland flow at a given time and this concept is known as partial area infiltration excess overland flow.

Another process responsible for overland flow generation is saturation excess overland flow which occurs when the data zone is completely saturated. As a result, even if rain is falling at a small rate, overland flow is generated because the data has no capacity to absorb any more water. Saturation excess flow may also occur when the surface data has very low permeability due to which saturation occurs at the catchment rapidly. During a storm the area of saturated data expands and contract, depending upon the rainfall intensity, and this phenomenon is known as dynamic contributing area concept. Note that data profile at a place may also be saturated due to flow coming from upstream area (on the surface or under the surface) on addition to rainfall. The partial area concept suggested by Betson is a generalization of the Horton concept.

It is also noted that sometimes there are pathways below the surface due to rock fractures or macro pores etc. which allow water to move through the data rapidly. At times, these results on large flow contribution to the river through sub-surface pathways and this is known as sub-surface storm flow.

Factors affecting Runoff from a Catchment

Runoff from a catchment depends upon a number of factors. Major factors are discussed next.

Topography and Orientation

Topography of a catchment which has evolved over millions of years due to geomorphological processes plays an important role on runoff generation. on conjunction with other factors, topography also determines the flow pathways, storages, and types and density of vegetation. These ultimately influence catchment water and sediment yield. on catchment with steep slopes, water stays on the surface for a shorter time, the opportunity for infiltration is less and runoff is more if all other things are the same. Steep sloped catchments also have a shorter time of concentration. The average slope of river is an important parameter on some empirical methods to determine the time of concentration. If the catchment hillslopes are steep, water swiftly moves over the land, whereas on relatively flat catchments, more water infiltrates and runoff moves slowly. The same behavior is expected for the river network.

The aspect or orientation of the catchments is important particularly on the case of snow and glacier fed streams. North facing hillslopes receive less sunshine on the northern hemisphere and this results on more melt water to the streams and also more ET. Further, the orientation of the valley with respect to wind and storm direction also influences hydrologic processes on a catchment. Some models explicitly account for the topography of catchment through topography-based indices.

For example, the TOPMODEL uses a topographical index as a measure of flow accumulation at a given point on the catchment surface. It was first defined as a wetness index:

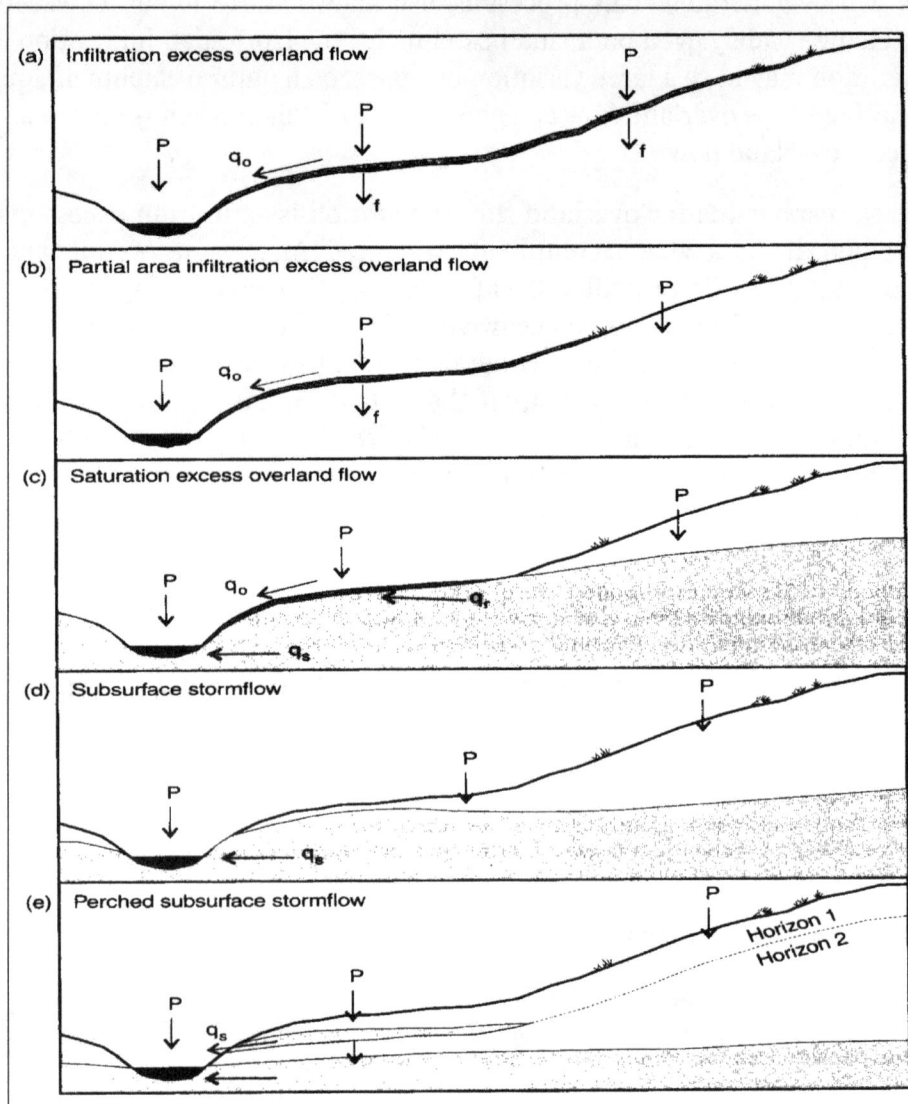

(a) Infiltration excess overland flow

(b) Partial area infiltration excess overland flow

(c) Saturation excess overland flow

(d) Subsurface stormflow

(e) Perched subsurface stormflow

Various hillslope runoff mechanisms.

$$\lambda = In \frac{\alpha}{\tan \beta}$$

where λ is the topographical wetness index, α is the local upslope area draining through a certain point per unit contour length and $\tan \beta$ is the local slope on radians. Topographical index controls the flow accumulation and data moisture.

Size and Shape

The size of a catchment naturally determines the magnitude of runoff generated due to several reasons. Usually the amount of water received by precipitation increases with size. The rational formula relates peak discharge from a catchment directly with area. However, on a large catchment, rainfall may have large variations and if the average rainfall decreases as we move downstream, the runoff per unit of catchment area may actually decrease. For example, on Krishna River basin on India, the head waters region receives very high rainfall which gradually decreases on the downstream direction. As a result, the runoff generated per unit area decreases with the increase on the catchment area. The time base of the hydrograph typically increases with catchment size.

The shape of the watershed controls the synchronization of flow from various parts to the outlet. Catchments with circular or fan shape produce runoff at higher rates compared to those which are elongated because runoff from different locations will reach the outlet at nearly the same time. Hydrographs from circular catchments will have sharper peaks and shorter time base. Elongated catchments produce hydrographs with smaller peaks and longer time base. The shape of river valleys also affects the flow velocity and hence the time to peak.

Soil and Geology

Data properties chiefly influence runoff on two ways. Infiltration of water depends upon data properties and thickness. The datas which are highly permeable will permit higher infiltration and less runoff. The depth of data profile determines the data moisture storage capacity. Deeper datas will absorb more water before attaining saturation compared to shallow datas. The properties of datas on a catchment also determine the type of vegetation and crops grown. Land use and vegetation impact catchment response. More vegetation will produce more resistance to overland flow and more infiltration (partly due to plant roots).

Whereas the geographical features, such as mountains and continents, have evolved over long time horizons, small-scale features, such as gullies and drainage channels, form over shorter time scales. Weathering, erosion and deposition are the main processes involved. The main processes that control short-term changes are weathering, erosion, transport, and deposition. Fluvial geomorphology is the study of the formation of rivers and how they respond to anthropogenic and climate induced changes.

Geological properties, such as types of rocks, fractures, and faults, also affect the occurrence and movement of sub-surface water. Some rock types permit easy storage and movement of water and such basins are rich on groundwater occurrence. The chemical composition of rocks also determines the quality of ground water as many minerals may get dissolved when water passes through these rocks.

Elements of the Hydrograph

A streamflow hydrograph is a graph of the time distribution of water discharge at a location. The graph is plotted with discharge on the ordinate and time on the abscissa. A hydrograph for a given storm reflects the influence of all the physical characteristics of the drainage basin and, to some extent, also reflects the characteristics of the storm causing the hydrograph. A hydrograph can be considered a thumbprint of the drainage basin. The actual shape of a hydrograph is determined by the rate at which water is transmitted from the various parts of the drainage basin to the outlet. Most of this water is carried by the channels, but some water flows overland directly to the outlet. Two drainage basins will not produce identical hydrographs for the same storm. Similarly, no two storms produce identical hydrographs from the same basin.

A number of conceptual models are available to describe runoff generation on different catchments or at different locations on the same catchment but at different times. Based on the path taken by water, streamflow may be divided into surface flow, interflow, and base flow. Figure shows the components of a typical hydrograph.

The streamflow hydrograph rises from the beginning of wet season and with increasing catchment area at the gauging site for two reasons: a) as we move downstream on a basin, the area contributing to flow increases and so does the amount of river flow; and b) the accumulative rainfall increase as the wet season (or storm) progresses and also the river flow. When the wet or rain season is over, we enter the falling limb of the hydrograph and flow begins to gradually recede. Hydrograph on figure corresponds to uniform rainfall and stationary storm. If rainfall is not uniform, hydrograph will have kinks. If storm is moving, the hydrograph shape might be different depending upon the direction of storm movement.

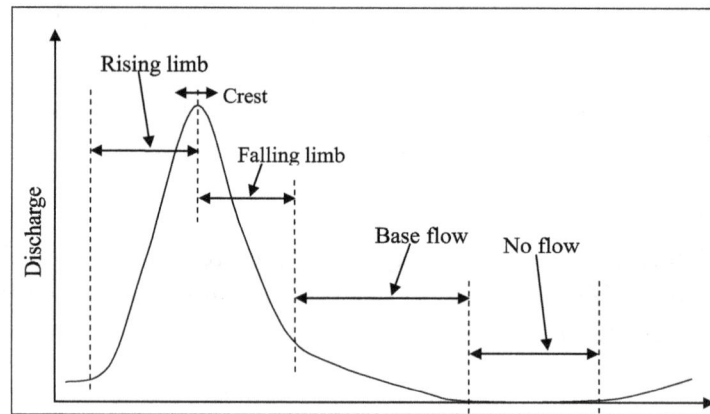

Components of a streamflow hydrograph.

Rising Limb

As surface runoff reaches the river, the water level on it begins to rise. As rain continues, more and more surface runoff reaches the river from larger areas, flow and water level continues to rise. After a certain time when rain stops or its intensity or areal coverage decreases, lesser amount of water reaches the channel and the discharge and water level begins to fall. The rising portion of the hydrograph is called the rising limb.

Crest

The time interval bracketing the highest discharge near the peak of the hydrograph is called the crest. on the case of a sharp peak, the crest will be of short duration, while for a flat peak, the crest segment covers a fairly long time interval. The crest represents a subjective zone of nearly equal highest discharges. The greatest discharge within the crest is the peak discharge, which is of primary interest on hydrologic design.

Recession Limb

Recession limb is the portion of the hydrograph after the crest segment. It is also known as the falling limb or the recession curve. The recession limb represents decreasing discharge as water drains out from the catchment storage after rainfall stops. The slope of the recession limb indicates the rate at which water is drained from the basin. The lower part of the recession after the inflexion point, which has a much lower slope, is believed to represent groundwater contribution because here water is withdrawn more slowly.

Streamflow recession can be expressed by an exponential decay:

$$Q_t = Q_0 K_r^t$$

where Q_0 is the initial discharge at any time, Q_t is the discharge at time interval t later, and K_r is the recession or depletion constant r dependent upon the unit of time and is less than unity.

Hydrograph Time Characteristics

Some critical characteristics and the shape of a hydrograph can be expressed on terms of a few time parameters.

- Time to Peak: The time to peak of a hydrograph is the time elapsed from the beginning of the rising limb to the peak discharge. This time depends on the drainage-basin characteristics, such as shape (elongated, circular, any other), distance from the most upstream point to the outlet, drainage density, channel slope and roughness, and data characteristics. Time to peak somewhat depends upon the distribution of rainfall over the basin. For a given amount of runoff, a longer time to peak has a lower peak discharge than a shorter time to peak.

- Time of Concentration: It is the time required for a drop of water which falls on the most remote part of the catchment to reach the outlet. If a rainfall of sufficient intensity continues for the time of concentration then the entire drainage basin would be contributing to the hydrograph at this time and the discharge will be the maximum that can occur from a given storm intensity over the catchment. Assuming that the rainfall is uniform over the entire catchment, the discharge increases as water from progressively farther distances arrives at the outlet. At and after the time of concentration, the discharge becomes constant since the entire basin is contributing to the flow.

One of the most commonly used formulas to compute the time of concentration was developed by Kirpich:

$$t_c = 0.0078\left(L / S^{0.5}\right)^{0.77}$$

where t_c is the time of concentration on minutes, L is the length of travel on feet from the most remote point on the drainage basin along the drainage channel to the basin outlet, and S is the slope on feet per foot (it can be determined by the difference on elevation of the most remote point and that of the outlet divided by L). The equation assumes uniform rainfall over the catchment.

Components of Streamflow

Figure shows a streamflow hydrograph and its components. The two main components of runoff are: (a) direct runoff and (b) baseflow. The direct runoff is divided into surface runoff and quick interflow, whereas the baseflow is divided into delayed interflow and groundwater runoff. The division into quick and delayed interflows is essentially arbitrary. The total runoff corresponds to a given storm event and its volume is determined by including on the streamflow hydrograph all runoff between the baseflow discharge occurring prior to the storm up to the same baseflow discharge after the storm. Verification of the pathways of water movement can be accomplished by isotopic techniques or by employing a processbased hydrologic model.

Surface Runoff

Surface runoff or overland flow is that water which travels over the ground surface to a drainage channel. Most surface runoff flows to first-order channels because they collectively drain the greatest area of the drainage basin. Surface runoff also includes that precipitation that falls directly on water flowing on the channel. Sheet flow usually occurs from an impervious surface such as a paved parking lot, but can only occur on a natural drainage basin when rainfall intensity uniformly exceeds the infiltration capacity. This condition does not frequently happen. Variations on the distribution of data type and of rainfall over a drainage basin usually result on limited sheet flow. Surface runoff is believed to be the principal contributor to the peak discharge from a storm event. Because this water runs off over the surface to the channel, it is the first to reach the channel and, hence, forms the rising limb and peak of the hydrograph.

As the catchment area at a gauging site increases, the slope of rising limb of the surface runoff hydrograph becomes flatter; typically discharge is a power function of area.

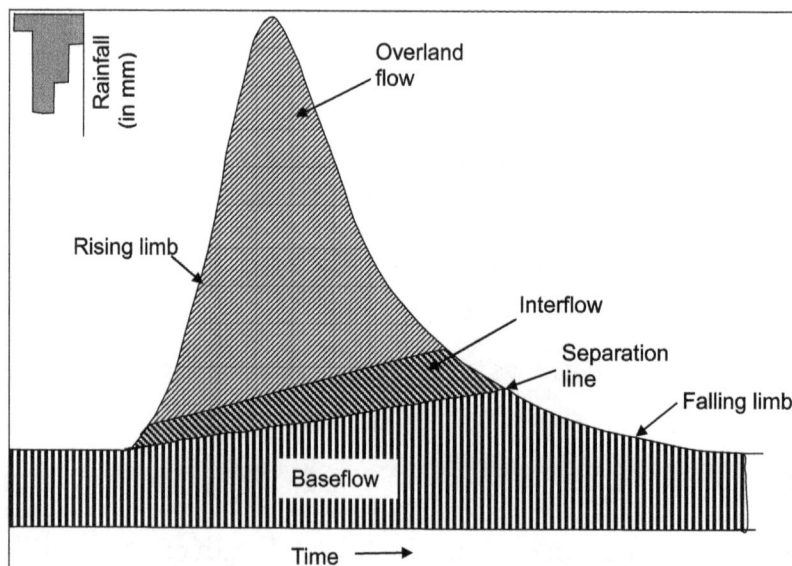

A typical streamflow hydrograph.

Interflow

Interflow, also called subsurface storm flow, is that surface water that infiltrates the surface layer and moves laterally beneath the surface to a channel. Interflow can occur on forest floors, where the leaves, needles, and other debris cover the ground. Interflow might occur on shallow datas filled and loosened by tree roots, rock debris covering the ground surface, or surface datas loosened by any cause. During interflow, the movement of water is subject to greater flow resistance than surface runoff. As a result, interflow does not move as rapidly as surface runoff. Accordingly, interflow does not add to the peak discharge, but reaches the outlet after the peak discharge has passed.

Direct Runoff

Direct runoff is usually considered to be the sum of surface runoff and interflow. Direct runoff is frequently equated with surface runoff. These two flow components move more rapidly than groundwater flow and for this reason are often lumped together for hydrologic purposes. Such lumping is reasonable for certain purposes because it is logical to believe that some interflow near the outlet will arrive at that point before surface runoff from farther up the basin.

Baseflow

Flow on a perennial stream prior to a storm is from baseflow. During a storm event, the baseflow is augmented by infiltration. Drainage basins with highly permeable, thick datas usually have a high groundwater-flow component and relatively small direct-flow component, whereas basins with heavy-clay, low-infiltration data have a small or zero groundwater component and a high direct-runoff component. A portion of the groundwaterflow component occurs from water infiltrating the banks of the channel during high-water flows.

ASCE defined base flow (or groundwater flow, fair weather flow) as the runoff that has reached the stream or river by passing first through the underlying aquifer, rather than by flowing directly on the ground surface. Hence, base flow is that part of river flow that is gradually entering the river from groundwater storage. Baseflow is also called as the groundwater flow, low flow, and fair weather flow.

Perennial streams depend on base flow for discharge between runoff producing events. The presence of base flow around the year indicates humid climate and a shallow water table that is hydraulically connected with the stream. Base flow is absent on (semi) arid climates and areas of deep groundwater. Contribution of base flow to streamflow depends upon the hydro-geology of the aquifers. Sources, such as lakes, marshes, snow, and river banks may also supply water for base flow. At a given time, more than one source may be supporting base flow. The movement of groundwater on the lateral direction is slower than vertical, since the hydraulic gradient for lateral movement is smaller. The supply of water from the aquifer to the river continues as long as there is adequate gradient.

As water moves to the stream from higher elevations, the hydraulic gradient decreases and lesser water will travel to the stream with time, unless there is additional infiltration to the aquifer. The gradual reduction of baseflow is called baseflow recession. Many different forms of base flow recession curves have been proposed. Among these, the most common is the exponential recession:

$$Q_t = Q_0 k_r^t$$

where Q_0 is the flow at time t = 0; Q_t is the flow at time t; and parameter k_r is known as the recession constant. It is a dimensionless quantity which depends upon the unit of time selected. It lies on the interval [0,1] and is normally greater than 0.7. Horton suggested a non-linear recession equation, known as the Horton's double exponential:

$$Q_t = Q_0 \exp\left(-a_2\, t^m\right)$$

where a_2 and m are constants.

To determine the recession constant k_r, one can plot Qt-1 versus Q_t on a simple graph. This plot will result on a straight line passing through the origin whose slope will be k_r.

The master baseflow recession curve is a composite recession curve which represents the mean recession behavior. Of course, the information about the recession variability is lost on the process. This curve can be constructed by observing discharge at a number of time intervals encompassing the entire dry-weather period. These are then plotted against time on a semilog paper, and a best-fit straight line is drawn through the plotted points. The resulting curve is the master baseflow recession curve.

Baseflow Separation

Baseflow separation is the process of separating surface runoff from baseflow. Even though such separation is somewhat arbitrary and subjective, it is useful on many analyses. Several techniques have been developed to perform base flow separation. The area method of base flow separation is based upon a nonlinear relation between time and area:

$$N = b\, A^{0.2}$$

where A is the drainage-basin area on km^2 ; b is a coefficient, equal to 0.8; and N is the time on days from the hydrograph peak. This equation is not suitable for smaller watersheds. It generally gives a longer time base. For example, if A = 1000 km^2, then N = 3.18 days, i.e., if rainfall occurs for 6 hours, its effect will be felt for more than 3 days.

It is convenient to draw a separation line directly from the chosen groundwater discharge on the receding limb to the point under the hydrograph peak. Although this linear separation does not represent the true boundary between direct runoff and groundwater runoff, the error may be acceptable on most cases.

The three-component separation involves separating surface runoff, interflow, and base flow. A method, developed by Barnes is illustrated on figure. First, streamflow recession is plotted on a semi-logarithmic paper. In figure, the groundwater recession plots approximately as a straight line, with K_r = 0.992. By extending this straight line under the hydrograph to the point directly under the point of inflection E and to B on line AB, points B and J are connected arbitrarily by a straight line. The area under the hydrograph above BJH is considered to be direct flow and that area below BJH is considered to be groundwater flow. The direct runoff is replotted and a straight line IL with K_r = 0.966 is fitted and extended to point I directly under the inflection point E and to the beginning point M. The line MIL divides the replotted hydrograph into surface runoff on the top and interflow below.

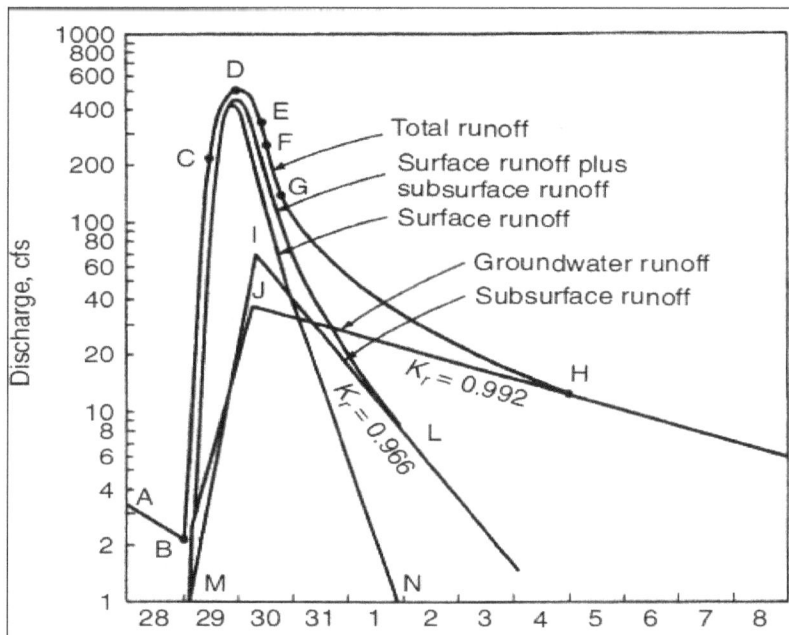

Three-component hydrograph separation.

The method developed by Singh and Stall is also followed by many engineers. Digital filters and chemical and isotope tracers provide another means to separate hydrograph on components.

Hydrological Measurements

Hydrological measurements are essential for the interpretation of water quality data and for water resource management. Variations on hydrological conditions have important effects on water quality. on rivers, such factors as the discharge (volume of water passing through a cross-section of the river on a unit of time), the velocity of flow, turbulence and depth will influence water quality. For example, the water on a stream that is on flood and experiencing extreme turbulence is likely to be of poorer quality than when the stream is flowing under quiescent conditions. This is clearly illustrated by the example of the hysteresis effect on river suspended sediments during storm events. Discharge estimates are also essential when calculating pollutant fluxes, such as where rivers cross international boundaries or enter the sea. on lakes, the residence time, depth and stratification are the main factors influencing water quality. A deep lake with a long residence time and a stratified water column is more likely to have anoxic conditions at the bottom than will a small lake with a short residence time and an unstratified water column.

It is important that personnel engaged on hydrological or water quality measurements are familiar, on general terms, with the principles and techniques employed by each other.

Rivers

Proper interpretation of the significance of water quality variables on a sample taken from a river requires knowledge of the discharge of the river at the time and place of sampling. on order to calculate the mass flux of chemicals on the water (the mass of a chemical variable passing a cross-section of the river on a unit of time), a time series of discharge measurement is essential.

The flow rate or discharge of a river is the volume of water flowing through a cross-section on a unit of time and is usually expressed as $m^3 s^{-1}$. It is calculated as the product of average velocity and cross-sectional area but is affected by water depth, alignment of the channel, gradient and roughness of the river bed. Discharge may be estimated by the slope-area method, using these factors on one of the variations of the Chezy equation. The simplest of the several variations is the Manning equation which, although developed for conditions of uniform flow on open channels, may give an adequate estimate of the non-uniform flow which is usual on natural channels. The Manning equation states that:

$$Q = \frac{1}{n} AR^{2/3} S^{1/2}$$

where,

      Q = discharge ($m^3 s^{-1}$).

      A = cross-sectional area (m^2).

      R = hydraulic radius (m) and = A/P.

      P = wetted perimeter (m).

      S = slope of gradient of the stream bed.

      n = roughness coefficient.

More accurate values for discharge can be obtained when a permanent gauging station has been established on a stretch of a river where there is a stable relationship between stage (water level) and discharge, and this has been measured and recorded. Once this relationship is established, readings need only be taken of stage, because the discharge may then be read from a stage-discharge curve.

Water quality samples do not have to be taken exactly at a gauging station. They may be taken a short distance upstream or downstream, provided that no significant inflow or outflow occurs between the sampling and gauging stations. The recommended distance is such that the area of the river basin upstream of the sampling station is between 95 per cent and 105 per cent of the area of the river basin upstream of the gauging station.

The stage-discharge relationship, or rating curve, usually includes the extremes of discharge encountered on a normal year. The rating curve should be checked periodically, ideally once a year, since minor adjustments may be necessary to take account of changes on the crosssection of the stream or instability on the flow characteristics, or to eliminate errors on previous measurements. Systematic variation as a result of unstable flow may be apparent if the stage-discharge relationship for a single flood event is examined. The discharge during the rising phase of a flood event is usually greater than that during the falling phase of the same flood event. While unstable flow can produce a loop on the stage-discharge plot for an individual storm event, it is not usually apparent on the annual rating curve that is commonly used on hydrological survey programmes. Unstable cross-sections cause stage-discharge variability and can produce sudden and significant shifts on the rating curve as a result of erosion or deposition of material on the river bed.

Measuring Stream Flow

If samples are to be taken at a point where the stage-discharge relationship is either unknown or unstable, discharge should be measured at the time of sampling. Discharge measurement, or stream gauging, requires special equipment and, sometimes, special installations. The measurements should be made by an agency that has staff with expertise on the techniques of hydrological survey. The most accurate method is to measure the crosssectional area of the stream and then, using a current meter, determine the average velocity on the cross-section. If a current meter is not available, a rough estimate of velocity can be made by measuring the time required for a weighted float to travel a fixed distance along the stream.

For best results, the cross-section of the stream at the point of measurement should have the following ideal characteristics:

- The velocities at all points are parallel to one another and at right angles to the crosssection of the stream.

- The curves of distribution of velocity on the section are regular on the horizontal and vertical planes.

- The cross-section should be located at a point where the stream is nominally straight for at least 50 m above and below the measuring station.

- The velocities are greater than 10-15 cm s^{-1}.

- The bed of the channel is regular and stable.

- The depth of flow is greater than 30 cm.

- The stream does not overflow its banks.

- There is no aquatic growth on the channel.

It is rare for all these characteristics to be present at any one measuring site and compromises usually have to be made.

Velocity varies approximately as a parabola from zero at the channel bottom to a maximum near the surface. A typical vertical velocity profile is shown in figure. It has been determined empirically that for most channels the velocity at six-tenths of the total depth below the surface is a close approximation to the mean velocity at that vertical line. However, the average of the velocities at two-tenths and eight-tenths depth below the surface on the same vertical line provides a more accurate value of mean velocity at that vertical line.

Velocity also varies across a channel, and measurements must, therefore, be made at several points across the channel. The depth of the river varies across its width, so the usual practice is to divide the cross-section of the stream into a number of vertical sections as shown in figure and measure velocity at each of these. No section should include more than 10-20 per cent of the total discharge. Thus, between 5 and 10 vertical sections are typical, depending on the width of the stream.

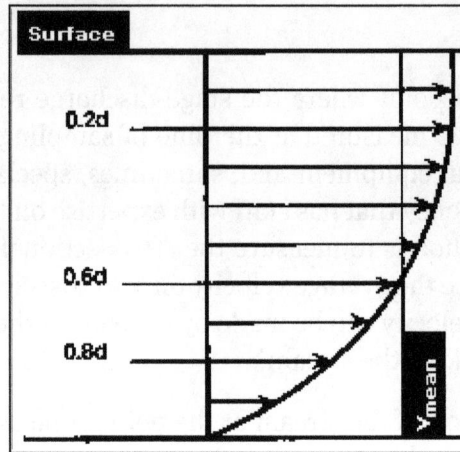

Typical river velocity profile on the vertical plane.

Procedure for measuring discharge:

- All measurements of distance should be made to the nearest centimetre.

- Measure the horizontal distance b1, from reference point 0 on shore to the point where the water meets the shore, point 1 in figure.

- Measure the horizontal distance b2 from reference point 0 to vertical line 2.

- Measure the channel depth d2 at vertical line 2.

- With the current meter make the measurements necessary to determine the mean velocity v2 at vertical line 2.

- Repeat steps 3, 4 and 5 at all the vertical lines across the width of the stream.

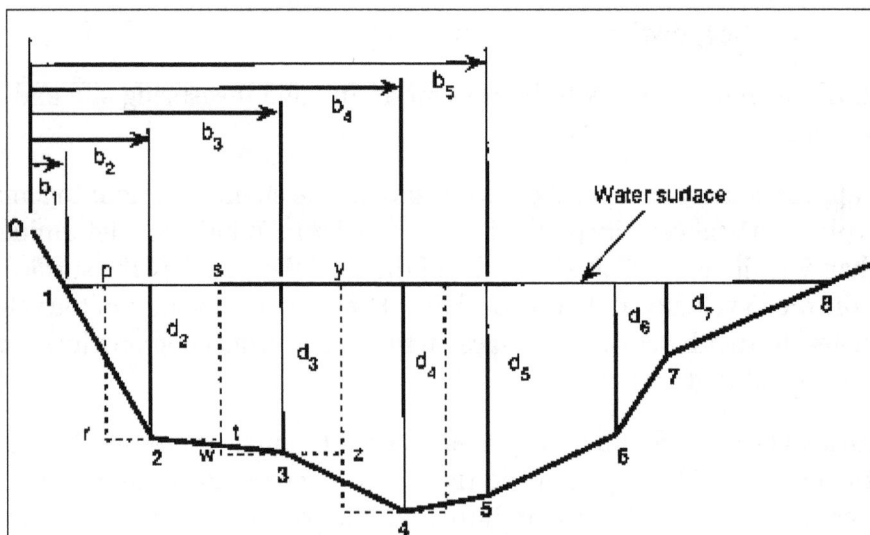

Cross-section of a stream divided into vertical sections for measurement of discharge.

The computation for discharge is based on the assumption that the average velocity measured at a vertical line is valid for a rectangle that extends half of the distance to the verticals on each side of it, as well as throughout the depth at the vertical. Thus, in figure, the mean velocity $\bar{v}_2$ would apply to a rectangle bounded by the dashed line p, r, s, t. The area of this rectangle is:

$$a_2 = \frac{b_3 - b_1}{2} \times d_2$$

and the discharge through it will be:

$$Q_2 = a_2 \times \bar{v}_2$$

Similarly, the velocity $\bar{v}_3$ applies to the rectangle s, w, z, y and the discharge through it will be:

$$Q_3 = \frac{b_4 - b_2}{2} \times d_3 \times \bar{v}_3$$

The discharge across the whole cross-section will be:

$$Q_T = Q_1 + Q_2 + Q_3 \dots Q_{(n-r)} + Q_n$$

In the example of figure, n = 8. The discharges on the small triangles at each end of the cross-section, Q_1 and Q_n, will be zero since the depths at points 1 and 8 are zero.

If the water is shallow, the operator may wade into the stream holding the current meter on place while measurements are being made. Where the water is too deep for wading (more than 1 metre) the current meter must be lowered from a bridge, an overhead cableway or a boat. The section where flow measurement is made does not have to be at exactly the same place as either the monitoring station or the water level indicator provided that there is no significant inflow or outflow between these points along the stream.

Bridges are preferred as stream gauging stations because they usually allow easy access to the full width of the stream, and a water level indicator can be fastened to a bridge pier or abutment. Aerial cableways are often located at places where characteristics of the stream cross-section approach the ideal. However, they necessitate a special installation, and this is often impractical for a water quality monitoring team. Velocity measurements made from a boat are liable to yield inaccurate results because any horizontal or vertical movement of the boat will be identified as velocity by the current meter. on shallow streams the water velocity close to the boat will be affected and this may distort the meter readings.

Floats should be used for velocity measurement only when it is impossible to use a current meter. A surface float will travel with a velocity about 1.2 times the mean velocity of the water column beneath it. A partially submerged float made from a wooden stick with a weight at its lower end (so that it floats vertically) may be used. The velocity of a float of this type will be closer to the mean velocity and a correction factor of about 1.1 is appropriate if the submerged part of the float is at one-third to one-half the water depth. The velocity of any float, whether on the surface or submerged, is likely to be affected by wind.

Discharge measurement may also be made by ultrasonic or electromagnetic methods, by injecting tracers into a stream, or by the construction of a weir. None of these methods is recommended because they either interfere with the natural water quality or they are prohibitively expensive and complex to install and operate.

Lakes and Reservoirs

In lakes and reservoirs, hydrological information (particularly water residence time) is needed for the interpretation of data and the management of water quality. These hydrological measurements are required on two different situations:

- When samples are to be taken from tributaries and outflowing streams.

- When samples are to be taken from the lake or reservoir itself.

Both types of sampling may aim at the estimation of the mass flow of some variable on the water body and, consequently, hydrological data are essential.

Sampling from Tributaries and Outflowing Streams

In outflowing streams, hydrological measurements should be obtained on the same manner as described above for rivers. on tributaries, the location of sampling stations and flow measurement stations should be selected so that backwater effects (water backing up the river from the lake) will be avoided. If this is not possible, the water level at the mouth of the tributary should be measured and recorded to provide data on the magnitude of the backwater and its variation with time.

Sampling from a Lake or Reservoir

The water level at the time of sampling must be measured. If the water surface is calm and a water level gauge has been installed, a single reading may be sufficient. If there is no official gauge, the water level should be recorded on relation to a conveniently located, identifiable point on a rock outcrop, large boulder or other landmark that is reasonably permanent. If there is any reason to suspect that this water level marker might move or be moved, reference should be made to a second landmark. The use of landmarks as a water level reference is a temporary measure and a water level gauge should be installed as soon as possible.

Waves and the inclination of the water surface (seiches) may cause problems on observing water levels. High waves may make it difficult for the observer to see the gauge and the continual motion of the water makes it impossible to determine the exact water level. on such conditions the observer should try to record the highest and lowest positions of the changing water level and calculate the average of the two. The wind condition should also be noted, together with an estimate of the height of the waves.

In certain conditions, current measurements on lakes or reservoirs provide information that is helpful on the interpretation of the results of analyses of water and sediment samples. Currents may cause water quality to vary appreciably within short distances or time periods. The flow velocities that normally occur on lakes are measured with sensitive recording current meters anchored at given depths. Sometimes, however, a rough estimate of the flow field (the general pattern of flow within a lake) can be made by observing the motion of surface floats. on reservoirs, the operation of valves or sluices can create localised currents which can affect the water quality on their vicinity.

Mass Flux Computation

Water quality monitoring often seeks to obtain two main types of estimations related to the physical and chemical variables being measured:

- The instantaneous values of the concentrations of water quality variables.

- The mass flux of water quality variables through river sections, lakes or reservoirs over specified periods of time.

For a river, if only one sample has been taken at a point, provided the concentration c of the variable of interest has been determined and the instantaneous discharge Q has been obtained, the instantaneous mass flux Q_m of that variable can be calculated from the formula $Q_m = cQ$.

When samples have been taken and discharge has been measured on each of the vertical sections across a stream, the instantaneous mass flux is given by:

$$\sum_{i=1}^{n} C_i \overline{V}_i A_i \quad gs^{-1}$$

Where,

       n = number of vertical sections.

       C_i = concentration of the variable on section i (mg l^{-1}).

       $\overline{V}_i$ = mean velocity on section i (m s^{-1}).

       A_i = cross-sectional area of section i (m^2).

The average concentration at the cross-section can be obtained from:

$$\frac{Q_m}{Q} \quad mg^{-1}$$

Where,

       Q = the instantaneous discharge ($m^3 s^{-1}$).

       Q_m = the instantaneous mass flux of the variable (g s^{-1}).

Calculation of the mass flux over a period (t_o ... t_m) may be determined quite accurately, but this requires many measurements of water quality and the use of a complex formula. Normally, water quality determinations are carried out at relatively long time intervals (weekly or monthly). By contrast, discharges are often determined daily, based on daily observation of the water level and the stage-discharge relationship. The simplest way to estimate daily concentrations of a variable is to assume that each measured value of concentration is valid for half of the preceding and following intervals between the collection of samples. However, this assumption is valid only when variations on the concentration of a variable are small. It is usually necessary to use complicated interpolation techniques.

Regression techniques are appropriate if a reliable relationship can be established between the concentrations of the variable of interest and some other chemical or physical variable that is measured at frequent intervals. The most suitable variable will often be discharge (determined easily from water level). Thus the consistency of the relationship between the concentration of a variable and the discharge should always be checked. If the relationship is reasonably consistent, discharge can serve as the basis on which to estimate the mass flux of a variable. If it is inconsistent, however, some other relationship should be sought.

In small streams and during flood peaks, the discharge may vary considerably over a 24- hour period. A variation on the concentration of a variable on excess of an order of magnitude is typical, and both sampling and discharge measurements need to be frequent. If the daily maximum flow is two to three times the average flow, a 4-hour interval between samples is recommended. If the maximum is greater than three times the average, discharge should be measured and samples taken every hour. This information can be obtained during a shortterm pilot study.

Groundwater

Groundwater is important as a significant water resource, on its own right, and also because of its interaction with surface water. Groundwater recharges streams and rivers on some areas, while on others it is itself recharged by surface water.

The hydrogeological conditions of water-bearing rocks or aquifers have a significant influence on the quality of groundwater on much the same way that hydrological conditions influence surface water quality. The rate of flow through the aquifer, residence time, inflows and outflows from the aquifer all influence the groundwater quality.

As the rate of flow of groundwater is much lower than that of surface water, there is a significant risk that contaminants can build up on aquifers to the point where the water becomes unusable. This could force the abandonment of boreholes and result on a permanent reduction on the quantity of usable groundwater.

There are certain influences that are unique to groundwater quality. These are the nature of the aquifer, the presence or absence of contaminants on unsaturated layers above the aquifer, the presence of naturally occurring contaminants on the aquifer and the interaction of groundwater, surface water and contaminant movement.

Groundwater Flow

Detailed investigation of groundwater flows over a large area requires specialist knowledge and equipment and the manipulation of predictive formulae. Groundwater flow is threedimensional and, therefore, more difficult to predict than surface water. on addition, whereas surface water flow direction can be easily predicted by topographical survey, groundwater flow direction depends on aquifer type and hydraulic conditions on the aquifer and is difficult to assess without carrying out pump tests and tracer studies. Information on the direction of groundwater flow can be obtained by mapping out water levels on boreholes within the same aquifer. This gives an indication of the hydraulic gradient (or piezometric surface) and, thus, an idea of groundwater movement.

Groundwater flow information will assist on the prediction of contaminant movement on groundwater, on particular the spread and speed of movement of contaminants after a polluting event. However, this prediction is a complex procedure which is often inaccurate and is complicated further by the lack of knowledge of contaminant behaviour on groundwater.

Flows within aquifers on the medium scale may be assessed through tracer studies, which will indicate direction and rate of flow. The rate of movement of water into particular wells can be quite easily evaluated by pump tests. These tests will also provide information on the depression of groundwater level around a well during pumping.

The usual procedure for a field pump test is that water is pumped from a production borehole at a constant rate, which is controlled and measured. This may be done using a flow meter on the discharge pipe. The change on the level of the water table (or piezometric surface) around the borehole is monitored by measuring the change on depth to the water table at observation wells surrounding the borehole.

Pump tests provide valuable information for calculating aquifer properties which, on turn, provide information of importance to water quality, such as rate of movement of groundwater into boreholes, the area of the aquifer that will be exploited by the borehole and where land use restrictions may need to be applied.

Full inventories of boreholes should be prepared and should include information concerning the pumping depth, yield, aquifer transmissivity and storage coefficient. This will permit groundwater movement and water quality to be modelled. Pumping from adjacent boreholes will influence the yield and draw-down of a production borehole.

It is important to collect information on any changes of static water level. on some areas where initial intensive groundwater abstraction leads to a drop on the water table, there has been significant contamination of the unsaturated data layers overlying the aquifer. When abstraction was later decreased, the water level rose and became contaminated through the dissolution of contaminants held on the data profile. This may cause problems if water is withdrawn on the future and may also increase the risk to the rest of the aquifer, thus potentially reducing the quality of groundwater available.

Hydrologic Budget

The hydrologic budget consists of inflows, outflows, and storage as shown on the following equation:

> Inflow = Outflow +/- Changes on Storage

Inflows add water to the different parts of the hydrologic system, while outflows remove water. Storage is the retention of water by parts of the system. Because water movement is cyclical, an inflow for one part of the system is an outflow for another.

> Precipitation = Evapotranspiration + Total Runoff, where

> Total Runoff = Direct Runoff + Base flow (groundwater component of stream flow)

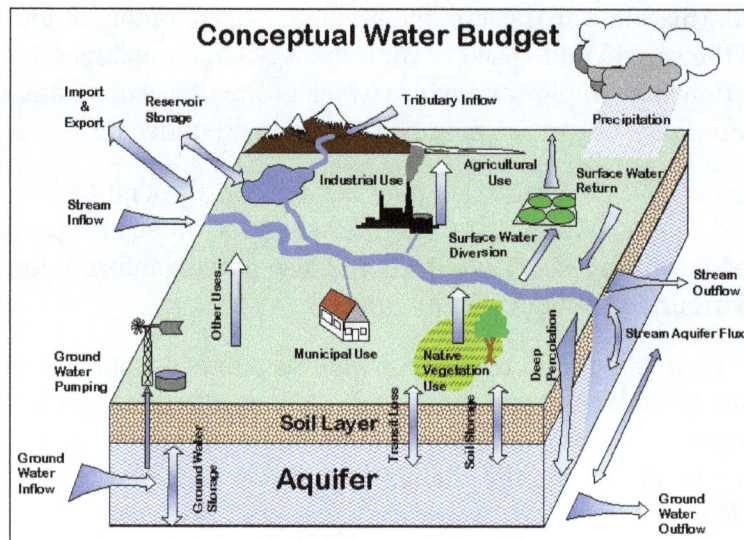

Hydrologic budget.

Looking at an aquifer as an example, percolation of water into the ground is an inflow to the aquifer. Discharge of groundwater from the aquifer to a stream is an outflow (also an inflow for the stream). Over time, if inflows to the aquifer are greater than its outflows, the amount of water stored on the aquifer will increase. Conversely, if the inflows to the aquifer are less than the outflows, the amount of water stored decreases. Inflows and outflows can occur naturally or result from human activity.

The earth's water supply remains constant, but man is capable of altering the cycle of that fixed supply. Population increases, rising living standards, and industrial and economic growth have placed greater demands on our natural environment. Our activities can create an imbalance on the hydrologic equation and can affect the quantity and quality of natural water resources available to current and future generations.

Water use by households, industries, and farms have increased. People demand clean water at reasonable costs, yet the amount of fresh water is limited and the easily accessible sources have been developed. As the population increases, so will our need to withdraw more water from rivers, lakes and aquifers, threatening local resources and future water supplies. A larger population will not only use more water but will discharge more wastewater. Domestic, agricultural, and industrial wastes, including the intensive use of pesticides, herbicides and fertilizers, often overload water supplies with hazardous chemicals and bacteria. Also, poor irrigation practices raise data salinity and evaporation rates. These factors contribute to a reduction on the availability of potable water, putting even greater pressure on existing water resources.

Large cities and urban sprawl particularly affect local climate and hydrology. Urbanization is accompanied by accelerated drainage of water through road drains and city sewer systems, which even increases the magnitude of urban flood events. This alters the rates of infiltration, evaporation, and transpiration that would otherwise occur on a natural setting. The replenishing of ground water aquifers does not occur or occurs at a slower rate.

Together, these various effects determine the amount of water on the system and can result on extremely negative consequences for river watersheds, lake levels, aquifers, and the environment as a whole. Therefore, it is vital to learn about and protect our water resources.

World Water Balance

Water balance is the most important integral physiographic characteristic of any territory—it determines its specific climate features, typical landscapes and opportunities for human land use. Assessment of mean long-term water balances of large regions at a sufficient accuracy depends on reliable estimation of the major water balance components—precipitation, evaporation and runoff (surface and subsurface).

The water balance of each continent (except Antarctica) is given separately for the areas of external runoff and internal runoff (endorheic areas) where precipitation is completely lost to evaporation. All balance components are estimated by independent methods which provide a computation of a balance discrepancy and thus assessment of reliability of the obtained results.

Areas of external runoff occupy about 80% of the Earth's land area. These areas receive 93% of precipitation onto the land; 88% of evaporation occurs there and 100% of the freshwater inflow to the World Ocean.

The whole land area (with islands) receives about 119 000 km^3 of precipitation during a year, or 800 mm. The maximum precipitation layer is observed on South America (1600 mm), the minimum precipitation layer (177 mm) occurs on Antarctica. On the other continents mean precipitation varies within 740 to 790 mm.

The total freshwater inflow to the World Ocean from the continents (without Antarctica) equals 39 500 km^3 /year. The oceans also receive about 2400 km^3 /year as subsurface runoff not drained by rivers, and 2300 km^3 of freshwater mainly as icebergs and melt runoff from the glaciers of Antarctica. Thus, the total freshwater inflow to the World Ocean from land is about 44 200 km^3 /year; this is equivalent to about 370 mm over the areas of external runoff or 300 mm, if related to the whole land area.

Evapotranspiration over continents varies from 420 mm to 850 mm (without Antarctica, where it is about zero). Evapotranspiration from the areas of external runoff is about 1.5 to 2 times greater than that from endorheic areas.

Evapotranspiration values given on the balance include runoff on endorheic areas where it is completely lost for evaporation. It also includes the amount of water used for human activities.

Analysis of water balance discrepancies shows good reliability of present assessment of long-term water balance for continents and large physiographic regions of the world.

Freshwater inflow to the World Ocean equals 502 000 km^3 /year, of which 91% is contributed by precipitation (458 000 km^3, or 1390 mm). More than a half of this amount is contributed to the Pacific Ocean.

The Arctic and Pacific Oceans have freshwater excess while the Atlantic and Indian Oceans have freshwater shortage. Freshwater excess is most significant on the southern areas of the Pacific, Atlantic and Indian Oceans because of the surplus of precipitation over evaporation there.

Evaporation from the surface of the World Ocean is about 1390 mm, varying from 220 mm from the Arctic Ocean to 1500 mm from the Pacific Ocean.

The highest freshwater inflow (49% of the total volume) from the continents occurs to the Atlantic Ocean. The effect of freshwater inflow is most important on the Arctic Ocean because its volume equals only 1.2% of the World Ocean volume.

Annual precipitation over the globe (numerically equal to evaporation) equals 577 000 km³, or 1130 mm. The depth of the evaporation layer from the ocean surface is three times greater than that from land. The volume of water annually evaporating from the ocean surface equals 87% and that from land is 13% of the total evapotranspiration from the Earth's surface.

Water balance is the ratio between water inflow and outflow estimated for different space and time scales, i.e., for the Earth as a whole, for oceans, continents, countries, natural-economic regions, and river basins, for a long-term period or for particular years and seasons. Water balance is the most important integral physiographic characteristic of any territory, determining its specific climate features, typical landscapes, possible water management and land use.

Analysis of water balance components for individual territories and time intervals is of great importance for studies of the hydrological cycle or water circulation on the atmosphere-hydrosphere-lithosphere system, as well as the underlying processes influenced by natural factors and human activities.

Precipitation, evaporation, river runoff and ground water outflow not drained by river systems are basic components determining water balance. Besides these components, there are minor components, too, e.g. moisture due to atmospheric water vapor condensation, deep artesian water outflow, or, conversely, recharge of deep aquifers, water losses for animal survival, etc. According to investigations, however, these components are very small if related to large river basins, regions and the globe—they are of no importance for water balance computation, so they can be ignored.

It should be noted that much fresh water is used on many regions for different human needs. Some of this is returned to water bodies as surface and subsurface runoff, but some water is lost, particularly to evaporation (from irrigated lands, reservoirs, etc.), thus increasing evapotranspiration on the region. This must be taken into account on the appropriate water balance components.

Thus, the assessments of water balances of large regions with sufficient accuracy is reduced to reliable determination of the main water balance components, i.e. precipitation, evaporation and runoff (surface and subsurface). Quantitative characteristics of these components for different regions of the Earth presented on this chapter are mainly based on the results of the global hydrological cycle studies carried out on Russia at the State Hydrological Institute (St Petersburg) and at the Institute of Water Problems (Moscow).

Water Balance Equations

Water balance equation for any land area and any time interval (without taking account of the above minor components) is as follows:

$$P + R'_s + R'_{un} = E + R_s + R_{un} \pm \Delta S$$

where, P is precipitation; E is evapotranspiration; R_s and R_{un}, R'_s and R'_{un} indicate surface and subsurface runoff from some land area and surface and subsurface water inflow to the land area, respectively; ΔS is water storage change on the area.

All terms on equation $P + R'_s + R'_{un} = E + R_s + R_{un} \pm \Delta S$, are on mm of water layer, which is a water volume for time unit divided by the area of the land.

If the water balance equation is considered for a long-term period it is simplified, because $\Delta S = 0$.

If the water balance is considered at the global scale, it should be noted that there are regions on each continent which differ greatly on their water balance structure. Most territories on the continents are the zones of so-called external runoff—river runoff from these zones discharges to the World Ocean directly. There are also rather large areas on the continents (probably except Antarctica) which have internal runoff. These endorheic areas are not connected to the World Ocean. River runoff formed on such regions is completely lost to evaporation.

For oceanic slopes and large river basins related to areas of external runoff from the continents (when it is possible to neglect surface and subsurface inflow from adjacent areas) the water balance equation for a long-term period is as follows:

$$P_{ext} = E_{ext} + R_{ext} + R_{un}$$

In equation $P_{ext} = E_{ext} + R_{ext} + R_{un}$, P_{ext} is a precipitation; R_{ext} is river runoff (from an oceanic slope) to the ocean (sea, lake); Run is subsurface runoff not drained by river systems and directly discharging to the ocean (sea, lake); E_{ext} is evapotranspiration including additional evaporation due to human ectivities.

In the areas of internal runoff (endorheic regions) the whole quantity of precipitation is ultimately lost to evaporation, so the water balance equation for a long-term period for such regions is as follows,

$$P_{int} = E_{int}$$

In equation $P_{int} = E_{int}$ P_{int} is precipitation; E_{int} is evapotranspiration from endorheic areas, including runoff formed within these areas and water losses for different human needs.

For a continent with available zones of external runoff and endorheic areas, the water balance equation would probably consist of equation $P_{ext} = E_{ext} + R_{ext} + R_{un}$ and equation $P_{int} = E_{int}$ joint together:

$$P_{ext} + P_{int} = E_{ext} + E_{int} + R_{ext} + R_{un}$$

For the World Ocean, as well as for individual oceans (and seas) the freshwater balance equation for the long-term period (without taking account of water exchange between the oceans) will be as follows:

$$E_{oc} = P_{oc} + R_{ext} + R_{un}$$

where: E_{oc} and P_{oc} are evaporation from the ocean and precipitation onto the ocean surface, respectively; R_{ext} and R_{un} are river water inflow and subsurface water inflow to the ocean.

For the whole Earth for a long-term period and a steady climate situation it is evident that the total precipitation should be equal to evaporation from the water surface plus evapotranspiration from land, i.e. for the world water balance the water balance equation similar to that for endorheic areas is valid:

$$P_{gl} = P_{oc} + P_{ext} + P_{int} = E_{oc} + E_{ext} + E_{int} = E_{gl}$$

where: P_{gl} and E_{gl} are global values of precipitation and evaporation from the Earth as a whole.

It should be noted that equation $P_{gl} = P_{oc} + P_{ext} + P_{int} = E_{oc} + E_{ext} + E_{int} = E_{gl}$, just like equations $P + R'_s + R'_{un} = E + R_s + R_{un} \pm \Delta S$ to $E_{oc} = P_{oc} + R_{ext} + R_{un}$, is valid if we assume that water coming from outer space is balanced by the amount of water vapour lost to space and deep water inflow (or juvenile water) is equivalent to water used for hydration of minerals on the lithosphere.

Applications of Hydrology in Engineering

Success of any water resources development project depends on timely and sufficient availability of water. Naturally proper assessment of this natural resource assumes great importance. By assessment we try to know on detail from where the resource comes, where it goes, at what time or when it comes and how much of it is really available.

Therefore, hydrological investigations form the first step on any water resources development scheme involving design, construction and operation of hydraulic structures. The history of hydraulic structures which have failed shows that majority of failures is due to insufficient hydrological analysis done while the structures were designed and constructed rather than due to structural weakness.

The cost of collecting sufficient hydrological data and its analysis constitutes an insignificant part of the total cost of the water resources development project but it ensures successful operation and life of the project and therefore becomes indispensible activity.

Although water is one of the most vital natural resources sometimes it brings destruction by way of storms and floods. An engineer is expected to forecast floods, to assure adequate storage capacity for irrigation, hydropower generation, industrial and domestic water supply, flood control etc.

The Practical Applications of the Knowledge of Hydrology are the following:

- Peak flow and future conditions of flow, at any point on the drainage valley can be correctly estimated for any basin or area.

- Spillway capacity can be accurately designed by estimating design flood.

- Design of river training work is facilitated.

- Dependable yields from the stream for generation of hydroelectric power can be calculated.

- Water supply to township and sewerage schemes can be properly designed.

- Water resources account of a river basin can be prepared.

- Reservoir capacity can be determined accurately.

- Operation of reservoirs can be done on an efficient manner.

Applications of Engineering Hydrology

- Hydrology is used to find out maximum probable flood at proposed sites e.g. Dams.

- The variation of water production from catchments can be calculated and described by hydrology.

- Engineering hydrology enables us to find out the relationship between a catchments's surface water and groundwater resources.

- The expected flood flows over a spillway, at a highway Culvert, or on an urban storm drainage system can be known by this very subject.

- It helps us to know the required reservoir capacity to assure adequate water for irrigation or municipal water supply on droughts condition.

- It tells us what hydrologic hardware (e.g. rain gauges, stream gauges etc) and software (computer models) are needed for real-time flood forecasting.

- Used on connection with design and operations of hydraulic structure.

- Used on prediction of flood over a spillway, at highway culvert or on urban storm drainage.

- Used to assess the reservoir capacity required to assure adequate water for irrigation or municipal water supply during drought.

- Hydrology is an indispensable tool on planning and building hydraulic structures.

- Hydrology is used for city water supply design which is based on catchments area, amount of rainfall, dry period, storage capacity, runoff evaporation and transpiration.

- Dam construction, reservoir capacity, spillway capacity, sizes of water supply pipelines and affect of afforest on water supply schemes, all are designed on basis of hydrological equations.

Hydraulic Engineering

Hydraulic Engineering is a specialised field within Environmental and Civil Engineering. Hydraulic systems are operated or fueled by the pressure of a fluid (e.g. water, oil etc.). Hydraulic Engineering deals with the technical challenges involved with water infrastructure and sewerage design. This discipline is really all about fluid flow and how it behaves on large quantities.

One main area of focus for Hydraulic Engineers is the design of water storage and transport facilities. Some examples include dams, channels, canals and lakes are all used to store and control water. Machinery which uses hydraulic power is also designed by Engineers on this discipline. Daily activities include designing structural elements that can withstand intense pressures.

Hydraulics

Hydraulics is mechanical function that operates through the force of liquid pressure. on hydraulics-based systems, mechanical movement is produced by contained, pumped liquid, typically through cylinders moving pistons. Hydraulics is a component mechatronics, which combines mechanical, electronics and software engineering on the designing and manufacturing of products and processes.

Simple hydraulic systems include aqueducts and irrigation systems that deliver water, using gravity to create water pressure. These systems essentially use water's own properties to make it deliver itself. More complex hydraulics use a pump to pressurize liquids (typically oils), moving a piston through a cylinder as well as valves to control the flow of oil.

A log splitter is a single-piston hydraulic machine that uses a valve at either end of the cylinder that allows the pistons to be moved by the pressurized liquid, driving a wedge to force wood into smaller pieces and return to a home position. Force multiplication can be created by using a cylinder with a smaller diameter to push a larger piston on a larger cylinder. Often, there will be a number of pistons. Industrial equipment such as backhoes often use a number of cylinders to move different parts. Electronic controls are generally used for these more complicated setups on large, powerful equipment.

Hydraulics are similar to pneumatic systems on function. Both systems use fluids but, unlike pneumatics, hydraulics use liquids rather than gasses. Hydraulics systems are capable of greater pressures: up to 10000 pounds per square inch (psi) vs about 100 psi on pneumatics systems. This pressure is due to the incompressibility of liquids which enables greater power transfer with increased efficiency as energy is not lost to compression, except on the case where air gets into hydraulic lines. Fluids used on hydraulics may lubricate, cool and transmit power as well. Pneumatics, being less multifaceted, require oil lubrication separately, which can be messy with air pressure. Pneumatics are simpler on design and to control, safer (with less risk of fire) and more reliable, partially as the compressibility of the gas-absorbing shock can protect the mechanism.

Fluid Mechanics

Fluid mechanics is the science concerned with the response of fluids to forces exerted upon them. It is a branch of classical physics with applications of great importance on hydraulic and aeronautical engineering, chemical engineering, meteorology, and zoology.

Basic Properties of Fluids

Fluids are not strictly continuous media on the way that all the successors of Euler and Bernoulli have assumed, for they are composed of discrete molecules. The molecules, however, are so small and, except on gases at very low pressures, the number of molecules per millilitre is so enormous that they need not be viewed as individual entities. There are a few liquids, known as liquid crystals, on which the molecules are packed together on such a way as to make the properties of the medium locally anisotropic, but the vast majority of fluids (including air and water) are isotropic. on fluid mechanics, the state of an isotropic fluid may be completely described by defining its mean mass per unit volume, or density (ρ), its temperature (T), and its velocity (v) at every point on space, and just what the connection is between these macroscopic properties and the positions and velocities of individual molecules is of no direct relevance.

A word perhaps is needed about the difference between gases and liquids, though the difference is easier to perceive than to describe. on gases the molecules are sufficiently far apart to move almost independently of one another, and gases tend to expand to fill any volume available to them. on liquids the molecules are more or less on contact, and the short-range attractive forces between them make them cohere; the molecules are moving too fast to settle down into the ordered arrays that are characteristic of solids, but not so fast that they can fly apart. Thus, samples of liquid can exist as drops or as jets with free surfaces, or they can sit on beakers constrained only by gravity, on a way that samples of gas cannot. Such samples may evaporate on time, as molecules one by one pick up enough speed to escape across the free surface and are not replaced. The lifetime of liquid drops and jets, however, is normally long enough for evaporation to be ignored.

There are two sorts of stress that may exist on any solid or fluid medium, and the difference between them may be illustrated by reference to a brick held between two hands. If the holder moves his hands toward each other, he exerts pressure on the brick; if he moves one hand toward his body and the other away from it, then he exerts what is called a shear stress. A solid substance such as a brick can withstand stresses of both types, but fluids, by definition, yield to shear stresses no matter how small these stresses may be. They do so at a rate determined by the fluid's viscosity. This property, about which more will be said later, is a measure of the friction that arises when adjacent layers of fluid slip over one another. It follows that the shear stresses are everywhere zero on a fluid at rest and on equilibrium, and from this it follows that the pressure (that is, force per unit area) acting perpendicular to all planes on the fluid is the same irrespective of their orientation (Pascal's law). For an isotropic fluid on equilibrium there is only one value of the local pressure (p) consistent with the stated values for ρ and T. These three quantities are linked together by what is called the equation of state for the fluid.

For gases at low pressures the equation of state is simple and well known. It is:

$$P = \left(\frac{RT}{M}\right)\rho,$$

where R is the universal gas constant (8.3 joules per degree Celsius per mole) and M is the molar mass, or an average molar mass if the gas is a mixture; for air, the appropriate average is about 29 $\times$ 10^{-3} kilogram per mole. For other fluids knowledge of the equation of state is often incomplete. Except under very extreme conditions, however, all one needs to know is how the density changes when the pressure is changed by a small amount, and this is described by the compressibility of the fluid—either the isothermal compressibility, β_T, or the adiabatic compressibility, β_S, according to circumstance. When an element of fluid is compressed, the work done on it tends to heat it up. If the heat has time to drain away to the surroundings and the temperature of the fluid remains essentially unchanged throughout, then β_T is the relevant quantity. If virtually none of the heat escapes, as is more commonly the case on flow problems because the thermal conductivity of most fluids is poor, then the flow is said to be adiabatic, and β_S is needed instead. (The S refers to entropy, which remains constant on an adiabatic process provided that it takes place slowly enough to be treated as "reversible" on the thermodynamic sense.) For gases that obey equation, it is evident that p and ρ are proportional to one another on an isothermal process, and:

$$\beta_T = \rho^{-1}\left(\frac{\partial\rho}{\partial p}\right)_T = p^{-1}$$

In reversible adiabatic processes for such gases, however, the temperature rises on compression at a rate such that:

$$T \propto \rho^{(\gamma-1)}, p \propto p^{\gamma}$$

and:

$$\beta_S = \rho^{-1}\left(\frac{\partial\rho}{\partial p}\right)_S = (\gamma p)^{-1} = \frac{\beta_T}{\gamma},$$

where γ is about 1.4 for air and takes similar values for other common gases. For liquids the ratio between the isothermal and adiabatic compressibilities is much closer to unity. For liquids, however, both compressibilities are normally much less than p^{-1}, and the simplifying assumption that they are zero is often justified.

The factor γ is not only the ratio between two compressibilities; it is also the ratio between two principal specific heats. The molar specific heat is the amount of heat required to raise the temperature of one mole through one degree. This is greater if the substance is allowed to expand as it is heated, and therefore to do work, than if its volume is fixed. The principal molar specific heats, CP and CV, refer to heating at constant pressure and constant volume, respectively, and:

$$\gamma = \frac{C_P}{C_V}.$$

For air, CP is about 3.5 R.

Solids can be stretched without breaking, and liquids, though not gases, can withstand stretching, too. Thus, if the pressure is steadily reduced on a specimen of very pure water, bubbles will ultimately appear, but they may not do so until the pressure is negative and well below -10^7 newton per square metre; this is 100 times greater on magnitude than the (positive) pressure exerted by the Earth's atmosphere. Water owes its high ideal strength to the fact that rupture involves breaking links of attraction between molecules on either side of the plane on which rupture occurs; work must be done to break these links. However, its strength is drastically reduced by anything that provides a nucleus at which the process known as cavitation (formation of vapour- or gas-filled cavities) can begin, and a liquid containing suspended dust particles or dissolved gases is liable to cavitate quite easily.

Work also must be done if a free liquid drop of spherical shape is to be drawn out into a long thin cylinder or deformed on any other way that increases its surface area. Here again work is needed to break intermolecular links. The surface of a liquid behaves, on fact, as if it were an elastic membrane under tension, except that the tension exerted by an elastic membrane increases when the membrane is stretched on a way that the tension exerted by a liquid surface does not. Surface tension is what causes liquids to rise up capillary tubes, what supports hanging liquid drops, what limits the formation of ripples on the surface of liquids, and so on.

Hydrostatics

It is common knowledge that the pressure of the atmosphere (about 10^5 newtons per square metre) is due to the weight of air above the Earth's surface, that this pressure falls as one climbs upward, and, correspondingly, that pressure increases as one dives deeper into a lake (or comparable body of water). Mathematically, the rate at which the pressure on a stationary fluid varies with height z on a vertical gravitational field of strength g is given by:

$$\frac{dp}{dz} = -\rho g.$$

If ρ and g are both independent of z, as is more or less the case on lakes, then:

$$p(z) = p(0) - \rho g z.$$

This means that, since ρ is about 10^3 kilograms per cubic metre for water and g is about 10 metres per second squared, the pressure is already twice the atmospheric value at a depth of 10 metres. Applied to the atmosphere, equation $p(z) = p(0) - \rho g z.$ would imply that the pressure falls to zero at a height of about 10 kilometres. on the atmosphere, however, the variation of ρ with z is far from negligible and $p(z) = p(0) - \rho g z.$ is unreliable as a consequence.

Differential Manometers

Instruments for comparing pressures are called differential manometers, and the simplest such instrument is a U-tube containing liquid, as shown in figure. The two pressures of interest, p_1 and p_2, are transmitted to the two ends of the liquid column through an inert gas—the density of which is negligible by comparison with the liquid density, ρ—and the difference of height, h, of the two menisci is measured. It is a consequence of $p(z) = p(0) - \rho g z$ that:

$$p_1 - p_2 = \rho g h.$$

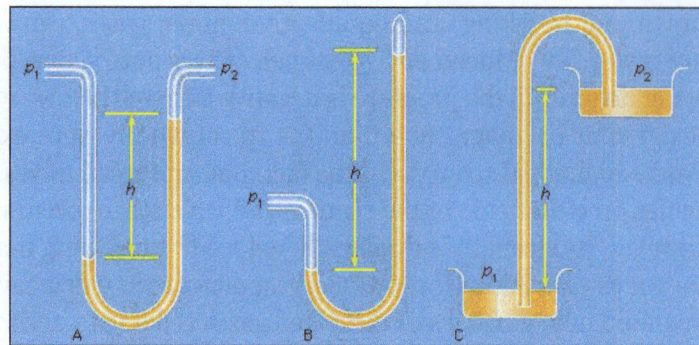

Schematic representations of (A) a differential manometer,
(B) a Torricellian barometer, and (C) a siphon.

A barometer for measuring the pressure of the atmosphere on absolute terms is simply a manometer on which p_2 is made zero, or as close to zero as is feasible. The barometer invented on the 17th century by the Italian physicist and mathematician Evangelista Torricelli, and still on use today, is a U-tube that is sealed at one end. It may be filled with liquid, with the sealed end downward, and then inverted. On inversion, a negative pressure may momentarily develop at the top of the liquid column if the column is long enough; however, cavitation normally occurs there and the column falls away from the sealed end of the tube, as shown on the figure. Between the two exists what Torricelli thought of as a vacuum, though it may be very far from that condition if the barometer has been filled without scrupulous precautions to ensure that all dissolved or adsorbed gases, which would otherwise collect on this space, have first been removed. Even if no contaminating gas is present, the Torricellian vacuum always contains the vapour of the liquid, and this exerts a pressure which may be small but is never quite zero. The liquid conventionally used on a Torricelli barometer is of course mercury, which has a low vapour pressure and a high density. The high density means that h is only about 760 millimetres; if water were used, it would have to be about 10 metres instead.

Figure illustrates the principle of the siphon. The top container is open to the atmosphere, and the pressure on it, p_2, is therefore atmospheric. To balance this and the weight of the liquid column on between, the pressure p_1 on the bottom container ought to be greater by $\rho g h$. If the bottom container is also open to the atmosphere, then equilibrium is clearly impossible; the weight of the liquid column prevails and causes the liquid to flow downward. The siphon operates only as long as the column is continuous; it fails if a bubble of gas collects on the tube or if cavitation occurs. Cavitation therefore limits the level differences over which siphons can be used, and it also limits (to about 10 metres) the depth of wells from which water can be pumped using suction alone.

Italian physicist and mathematician Evangelista Torricelli,
inventor of the mercury barometer.

Archimedes' Principle

Consider now a cube of side d totally immersed on liquid with its top and bottom faces horizontal. The pressure on the bottom face will be higher than on the top by $\rho g d$, and, since pressure is force per unit area and the area of a cube face is d^2, the resultant upthrust on the cube is $\rho g d^3$. This is a simple example of the so-called Archimedes' principle, which states that the upthrust experienced by a submerged or floating body is always equal to the weight of the liquid that the body displaces. As Archimedes must have realized, there is no need to prove this by detailed examination of the pressure difference between top and bottom. It is obviously true, whatever the body's shape. It is obvious because, if the solid body could somehow be removed and if the cavity thereby created could somehow be filled with more fluid instead, the whole system would still be on equilibrium. The extra fluid would, however, then be experiencing the upthrust previously experienced by the solid body, and it would not be on equilibrium unless this were just sufficient to balance its weight.

Archimedes' problem was to discover, by what would nowadays be called a nondestructive test, whether the crown of King Hieron II was made of pure gold or of gold diluted with silver. He understood that the pure metal and the alloy would differ on density and that he could determine the density of the crown by weighing it to find its mass and making a separate measurement of its volume. Perhaps the inspiration that struck him (in his bath) was that one can find the volume of any object by submerging it on liquid on something like a measuring cylinder (i.e., on a container with vertical sides that have been suitably graduated) and measuring the displacement of the liquid surface. If so, he no doubt realized soon afterward that a more elegant and more accurate method for determining density can be based on the principle that bears his name. This method involves weighing the object twice, first, when it is suspended on a vacuum (suspension on air will normally suffice) and, second, when it is totally submerged on a liquid of density ρ. If the density of the object is ρ', the ratio between the two weights must be:

$$\frac{W_2}{W_1} = \frac{(\rho' - \rho)}{\rho'}.$$

If ρ' is less than ρ, then W_2, according to equation $\frac{W_2}{W_1} = \frac{(\rho' - \rho)}{\rho'}$, is negative. What that means is that the object does not submerge of its own accord; it has to be pushed downward to make it do so. If an object with a mean density less than that of water is placed on a lake and not subjected to any downward force other than its own weight, it naturally floats on the surface, and Archimedes' principle shows that on equilibrium the volume of water which it displaces is a fraction ρ'/ρ of its own volume. A hydrometer is an object graduated on such a way that this fraction may be measured. By floating a hydrometer first on water of density ρ_0 and then on some other liquid of density ρ_1 and comparing the readings, one may determine the ratio ρ_1/ρ_0—i.e., the specific gravity of the other liquid.

$$\frac{W_2}{W_1} = \frac{(\rho' - \rho)}{\rho'}.$$

In what orientation an object floats is a matter of grave concern to those who design boats and those who travel on them. A simple example will suffice to illustrate the factors that determine orientation. Figure shows three of the many possible orientations that a uniform square prism might adopt when floating, with half its volume submerged on a liquid for which ρ = 2ρ′; they are separated by rotations of 22.5°. on each of these diagrams, C is the centre of mass of the prism, and B, a point known as the centre of buoyancy, is the centre of mass of the displaced water. The distributed forces acting on the prism are equivalent to its weight acting downward through C and to the equal weight of the displaced water acting upward through B. on general, therefore, the prism experiences a torque. In figure the torque is counterclockwise, and so it turns the prism away from 2A and toward 2C. on 2C the torque vanishes because B is now vertically below C, and this is the orientation that corresponds to stable equilibrium. The torque also vanishes on 2A, and the prism can on principle remain indefinitely on that orientation as well; the equilibrium on this case, however, is unstable, and the slightest disturbance will cause the prism to topple one way or the other. on fact, the potential energy of the system, which increases on a linear fashion with the difference on height between C and B, is at its smallest on orientation 2C and at its largest on orientation 2A. To improve the stability of a floating object one should, if possible, lower C relative to B. on the case of a boat, this may be done by redistributing the load inside.

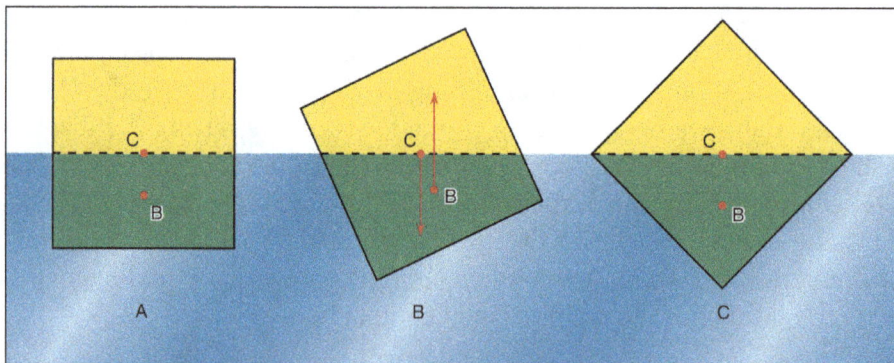

Three possible orientations of a uniform square prism floating on liquid of twice its density. The stable orientation is (C).

Surface Tension of Liquids

Of the many hydrostatic phenomena on which the surface tension of liquids plays a role, the most significant is probably capillarity. Consider what happens when a tube of narrow bore, often called a capillary tube, is dipped into a liquid. If the liquid "wets" the tube (with zero contact angle), the liquid surface inside the tube forms a concave meniscus, which is a virtually spherical surface having the same radius, r, as the inside of the tube. The tube experiences a downward force of magnitude $2\pi r d\sigma$, where σ is the surface tension of the liquid, and the liquid experiences a reaction of equal magnitude that lifts the meniscus through a height h such that:

$$2\pi r\sigma = \pi r^2 h\rho g$$

—i.e., until the upward force for which surface tension is responsible is balanced by the weight of the column of liquid that has been lifted. If the liquid does not wet the tube, the meniscus is convex and depressed through the same distance h. A simple method for determining surface tension involves the measurement of h on one or the other of these situations and the use of equation $2\pi r\sigma = \pi r^2 h\rho g$ thereafter.

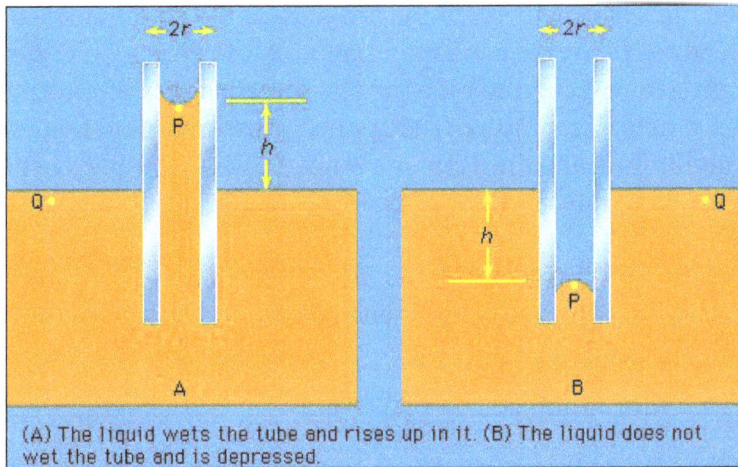

(A) The liquid wets the tube and rises up in it. (B) The liquid does not wet the tube and is depressed.

Capillarity.

It follows from equations $p(z) = p(0) - \rho gz$. and $2\pi r\sigma = \pi r^2 h\rho g$ that the pressure at a point P just below the meniscus differs from the pressure at Q by an amount:

$$\rho gh = \frac{2\sigma}{r};$$

it is less than the pressure at Q on the case to which figure refers and greater than the pressure at Q on the other case. Since the pressure at Q is just the atmospheric pressure, it is equal to the pressure at a point immediately above the meniscus. Hence, on both instances there is a pressure difference of $2\sigma/r$ between the two sides of the curved meniscus, and on both the higher pressure is on the inner side of the curve. Such a pressure difference is a requirement of equilibrium wherever a liquid surface is curved. If the surface is curved but not spherical, the pressure difference is:

$$\sigma\left(r_1^{-1} + r_2^{-1}\right),$$

where r_1 and r_2 are the two principal radii of curvature. If it is cylindrical, one of these radii is infinite, and, if it is curved on opposite directions, then for the purposes of they should be treated as being of $2\pi r\sigma = \pi r^2 h\rho g$ opposite sign.

The diagrams in figure were drawn to represent cross sections through cylindrical tubes, but they might equally well represent two vertical parallel plates that are partly submerged on the liquid a small distance apart. Consideration of how the pressure varies with height shows that over the range of height h the plates experience a greater pressure on their outer surfaces than on their inner surfaces; this is true whether the liquid wets both plates or not. It is a matter of observation that small objects floating near one another on the surface of a liquid tend to move toward one another, and it is the pressure difference just referred to that makes them behave on this way.

One other problem having to do with surface tension will be considered here. The diagrams in figure show stages on the growth of a liquid drop on the end of a tube which the liquid is supposed to wet. on passing from stage A to stage B, by which time the drop is roughly hemispheric on shape, the radius of curvature of the drop diminishes; and it follows from $\rho gh = \dfrac{2\sigma}{r}$ that, to bring about this growth, one must slowly increase the pressure of the liquid inside the tube. If the pressure could be held steady at the value corresponding to B, the drop would then become unstable, because any further growth (e.g., to the more or less spherical shape indicated in figure) would involve an increase on radius of curvature. The applied pressure would then exceed that required to hold the drop on equilibrium, and the drop would necessarily grow bigger still. on practice, however, it is easier to control the rate of flow of water through the tube, and hence the rate of growth of the drop, than it is to control the pressure. If the rate of flow is very small, drops will form the nonspherical shapes suggested by figure before they detach themselves and fall. It is not an easy matter to analyze the shape of a drop on the point of detachment, and there is no simple formula for the volume of the drop after it is detached.

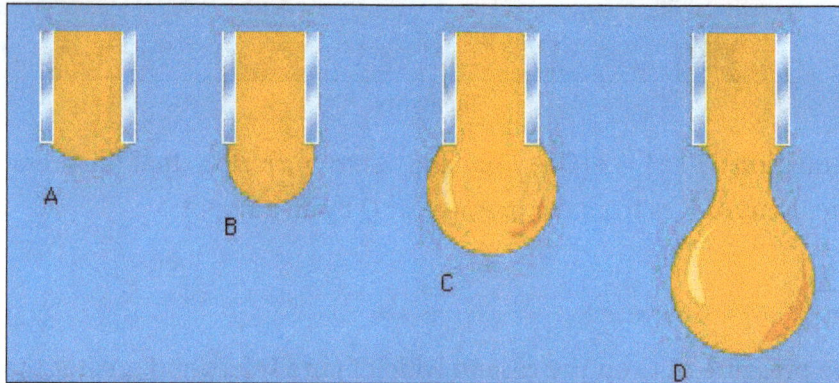

Stages on the formation of a liquid drop.

Hydrodynamics

Bernoulli's Law

Up to now the focus has been fluids at rest. This section deals with fluids that are on motion on a steady fashion such that the fluid velocity at each given point on space is not changing with time. Any flow pattern that is steady on this sense may be seen on terms of a set of streamlines, the trajectories of imaginary particles suspended on the fluid and carried along with it. on steady flow, the fluid is on motion but the streamlines are fixed. Where the streamlines crowd together, the fluid velocity is relatively high; where they open out, the fluid becomes relatively stagnant.

When Euler and Bernoulli were laying the foundations of hydrodynamics, they treated the fluid as an idealized inviscid substance on which, as on a fluid at rest on equilibrium, the shear stresses associated with viscosity are zero and the pressure p is isotropic. They arrived at a simple law relating the variation of p along a streamline to the variation of v (the principle is credited to Bernoulli, but Euler seems to have arrived at it first), which serves to explain many of the phenomena that real fluids on steady motion display. To the inevitable question of when and why it is justifiable to neglect viscosity, there is no single answer.

Consider a small element of fluid of mass m, which—apart from the force on it due to gravity—is acted on only by a pressure p. The latter is isotropic and does not vary with time but may vary from point to point on space. It is a well-known consequence of Newton's laws of motion that, when a particle of mass m moves under the influence of its weight mg and an additional force F from a point P where its speed is v_P and its height is z_P to a point Q where its speed is v_Q and its height is

z_Q, the work done by the additional force is equal to the increase on kinetic and potential energy of the particle—i.e., that:

$$\int_P^Q F \cdot ds = \left(\frac{1}{2}\right) m \left(v_Q^2 - v_P^2\right) + mg\left(z_Q - z_P\right).$$

In the case of the fluid element under consideration, F may be related on a simple fashion to the gradient of the pressure, and one finds:

$$\int_P^Q F \cdot ds = -m\int_P^Q \rho^{-1} dp.$$

If the variations of fluid density along the streamline from P to Q are negligibly small, the factor ρ^{-1} may be taken outside the integral on the right-hand side of, which thereupon reduces to $\rho^{-1}(p_Q - p_P)$. Then

$$\int_P^Q F \cdot ds = \left(\frac{1}{2}\right) m \left(v_Q^2 - v_P^2\right) + mg\left(z_Q - z_P\right) \text{ and } \int_P^Q F \cdot ds = -m\int_P^Q \rho^{-1} dp \text{ can be combined to obtain:}$$

$$\int_P^Q F \cdot ds = -m\int_P^Q \rho^{-1} dp.$$

$$\int_P^Q F \cdot ds = \left(\frac{1}{2}\right) m \left(v_Q^2 - v_P^2\right) + mg\left(z_Q - z_P\right).$$

$$\frac{P_P}{\rho} + \frac{v_P^2}{2} + gz_P = \frac{P_Q}{\rho} + \frac{v_Q^2}{2} + gz_Q.$$

Since this applies for any two points that can be visited by a single element of fluid, one can immediately deduce Bernoulli's (or Euler's) important result that along each streamline on the steady flow of an inviscid fluid the quantity:

$$\left(\frac{p}{\rho} + \frac{v^2}{2} + gz\right)$$

is constant.

Under what circumstances are variations on the density negligibly small? When they are very small compared with the density itself—i.e., when

$$\left(\frac{\Delta\rho}{\rho}\right) = \beta_s \Delta p = \left(\beta_s \rho\right)\Delta\left(\frac{v^2}{2} + gz\right) = \frac{\Delta\left(\frac{v^2}{2} + gz\right)}{V_s^2} \ll 1,$$

where the symbol Δ is used to represent the extent of the change along a streamline of the quantity that follows it, and where V_s is the speed of sound. This condition is satisfied for all the flow problems having to do with water that are discussed below. If the fluid is air, it is adequately satisfied provided that the largest excursion on z is on the order of metres rather than kilometres and provided that the fluid velocity is everywhere less than about 100 metres per second.

Bernoulli's law indicates that, if an inviscid fluid is flowing along a pipe of varying cross section, then the pressure is relatively low at constrictions where the velocity is high and relatively high where the pipe opens out and the fluid stagnates. Many people find this situation paradoxical when they first encounter it. Surely, they say, a constriction should increase the local pressure rather than diminish it? The paradox evaporates as one learns to think of the pressure changes along the pipe as cause and the velocity changes as effect, instead of the other way around; it is only because the pressure falls at a constriction that the pressure gradient upstream of the constriction has the right sign to make the fluid accelerate.

Paradoxical or not, predictions based on Bernoulli's law are well-verified by experiment. Try holding two sheets of paper so that they hang vertically two centimetres or so apart and blow downward so that there is a current of air between them. The sheets will be drawn together by the reduction on pressure associated with this current. Ships are drawn together for much the same reason if they are moving through the water on the same direction at the same speed with a small distance between them. on this case, the current results from the displacement of water by each ship's bow, which has to flow backward to fill the space created as the stern moves forward, and the current between the ships, to which they both contribute, is stronger than the current moving past their outer sides. As another simple experiment, listen to the hissing sound made by a tap that is almost, but not quite, turned off. What happens on this case is that the flow is so constricted and the velocity within the constriction so high that the pressure on the constriction is actually negative. Assisted by the dissolved gases that are normally present, the water cavitates as it passes through, and the noise that is heard is the sound of tiny bubbles collapsing as the water slows down and the pressure rises again on the other side.

Two practical devices that are used by hydraulic engineers to monitor the flow of liquids though pipes are based on Bernoulli's law. One is the venturi tube, a short length with a constriction on it of standard shape, which may be inserted into the pipe proper. If the velocity at point P, where the tube has a cross-sectional area A_P, is v_P and the velocity on the constriction, where the area is A_Q, is v_Q, the continuity condition—the condition that the mass flowing through the pipe per unit time has to be the same at all points along its length—suggests that $\rho_P A_P v_P = \rho_Q A_Q v_Q$, or that $A_P v_P = A_Q v_Q$ if the difference between ρ_P and ρ_Q is negligible. Then Bernoulli's law indicates:

$$\rho g h = \left(p_P - p_Q\right) = \left(\frac{1}{2}\right) \rho v_P^2 \left[\left(\frac{A_P}{A_Q}\right)^2 - 1\right].$$

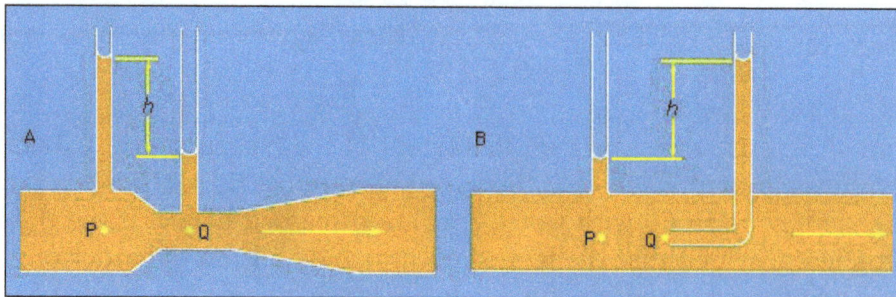

Schematic representation of (A) a venturi tube and of (B) a pitot tube.

Thus one should be able to find v_p, and hence the quantity $Q\,(= A_p v_p)$ that engineers refer to as the rate of discharge, by measuring the difference of level h of the fluid on the two side tubes shown on the diagram. At low velocities the pressure difference $(p_p - p_Q)$ is greatly affected by viscosity, and equation above is unreliable on consequence. The venturi tube is normally used, however, when the velocity is large enough for the flow to be turbulent. on such a circumstance, equation

$$\rho gh = \left(p_p - p_Q\right) = \left(\frac{1}{2}\right)\rho v_P^2 \left[\left(\frac{A_P}{A_Q}\right)^2 - 1\right]$$ predicts values for Q that agree with values measured by

more direct means to within a few parts percent, even though the flow pattern is not really steady at all.

$$\rho gh = \left(p_p - p_Q\right) = \left(\frac{1}{2}\right)\rho v_P^2 \left[\left(\frac{A_P}{A_Q}\right)^2 - 1\right].$$

The other device is the pitot tube, which is illustrated in figure. The fluid streamlines divide as they approach the blunt end of this tube, and at the point marked Q on the diagram there is complete stagnation, since the fluid at this point is moving neither up nor down nor to the right. It follows immediately from Bernoulli's law that:

$$\rho gh = \left(P_p - P_Q\right) = \left(\frac{1}{2}\right)\rho v_P^2.$$

As with the venturi tube, one should therefore be able to find v_p from the level difference h.

One other simple result deserves mention here. It concerns a jet of fluid emerging through a hole on the wall of a vessel filled with liquid under pressure. Observation of jets shows that after emerging they narrow slightly before settling down to a more or less uniform cross section known as the vena contracta. They do so because the streamlines are converging on the hole inside the vessel and are obliged to continue converging for a short while outside. It was Torricelli who first suggested that, if the pressure excess inside the vessel is generated by a head of liquid h, then the velocity v at the vena contracta is the velocity that a free particle would reach on falling through a height h—i.e., that

$$v = \sqrt{(2gh)}.$$

This result is an immediate consequence, for an inviscid fluid, of the principle of energy conservation that Bernoulli's law enshrines.

In the following section, Bernoulli's law is used on an indirect way to establish a formula for the speed at which disturbances travel over the surface of shallow water. The explanation of several interesting phenomena having to do with water waves is buried on this formula. This form of the law is restricted to gases on steady flow but is not restricted to flow velocities that are much less than the speed of sound. The complication that viscosity represents is again ignored throughout these two sections.

Waves on Shallow Water

Imagine a layer of water with a flat base that has a small step on its surface, dividing a region on which the depth of the water is uniformly equal to D from a region on which it is uniformly equal to $D(1 + \varepsilon)$, with $\varepsilon \ll 1$. Let the water on the shallower region flow toward the step with some uniform speed V, as figure suggests, and let this speed be just sufficient to hold the step on the same position so that the flow pattern is a steady one. The continuity condition (*i.e.*, the condition that as much water flows out to the left per unit time as flows on from the right) indicates that on the deeper region the speed of the water is $V(1 + \varepsilon)^{-1}$. Hence by applying Bernoulli's law to the points marked P and Q on the diagram, which lie on the same streamline and at both of which the pressure is atmospheric, one may deduce that:

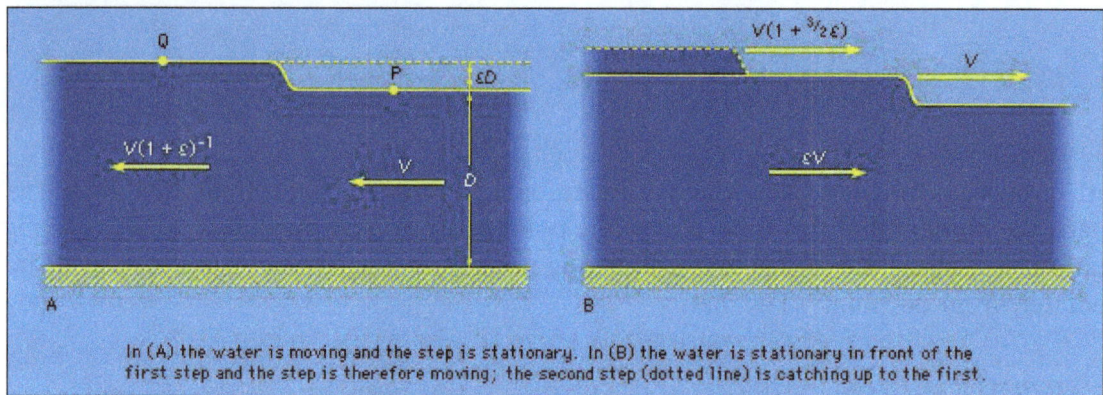

In (A) the water is moving and the step is stationary. In (B) the water is stationary in front of the first step and the step is therefore moving; the second step (dotted line) is catching up to the first.

Steps on the surface of shallow water.

$$g\varepsilon D = \left(\frac{1}{2}\right)V^2\left[1-(1+\varepsilon)^{-2}\right] \approx \varepsilon V^2$$

$-i.e,$

$$V = \sqrt{(gD)}.$$

This result shows that, if the water on the shallower region is on fact stationary, the step advances over it with the speed V that equation above describes, and it reveals incidentally that behind the step the deeper water follows up with speed $V[1 - (1 + \varepsilon)^{-1}] \approx \varepsilon V$. The argument may readily be extended to disturbances of the surface that are undulatory rather than steplike. Provided that the distance between successive crests—a distance known as the wavelength and denoted by λ—is much greater than the depth of the water, D, and provided that its amplitude is very much less than D, a wave travels over stationary water at a speed given by. Because their speed does not depend on wavelength, the waves are said to be nondispersive.

Evidently waves that are approaching a shelving beach should slow down as D diminishes. If they are approaching it at an angle, the slowing-down effect bends, or refracts, the wave crests so that they are nearly parallel to the shore by the time they ultimately break.

Suppose now that a small step of height εD ($\varepsilon \ll 1$) is traveling over stationary water of uniform depth D and that behind it is a second step of much the same height traveling on the same direction. Because the second step (suggested by a dotted line in figure) is traveling on a base that is moving at $\varepsilon\sqrt{(gD)}$ and because the thickness of that base is $(1 + \varepsilon)D$ rather than D, the speed of the second step is approximately $(1 + 3\varepsilon/2)\sqrt{(gD)}$. Since this is greater than $\sqrt{(gD)}$, the second step is bound to catch up with the first. Hence, if there are a succession of infinitesimal steps that raise the depth continuously from D to some value D', which differs significantly from D, then the ramp on the surface is bound to become steeper as it advances. It may be shown that if D' exceeds about 1.3D, the ramp ultimately becomes a vertical step of finite height and that the step then "breaks." A finite step that has broken dissipates energy as heat on the resultant foaming motion, and Bernoulli's equation is no longer applicable to it. A simple argument based on conservation of momentum rather than energy, however, suffices to show that its velocity of propagation is

$$\sqrt{\left(\frac{gD'(D'+D)}{2D}\right)}.$$

Tidal bores, which may be observed on some estuaries, are examples on the large scale of the sort of phenomena to which applies. Examples on a smaller scale include the hydraulic jumps that are commonly seen below weirs and sluice gates where a smooth stream of water suddenly rises at a foaming front. on this case, describes the speed of the water, since the front itself is more or less stationary.

When water is shallow but not extremely shallow, so that correction terms of the order of $(D/\lambda)^2$ are significant, waves of small amplitude become slightly dispersive . on this case, a localized disturbance on the surface of a river or canal, which is guided by the banks on such a way that it can propagate on one direction only, is liable to spread as it propagates. If its amplitude is not small, however, the tendency to spread due to dispersion may on special circumstances be subtly balanced by the factors that cause waves of relatively large amplitude to form bores, and the result is a localized hump on the surface, of symmetrical shape, which does not spread at all. The phenomenon was first observed on a canal near Edinburgh on 1834 by a Scottish engineer named Scott Russell; he later wrote a graphic account of following on horseback, for well over a kilometre, a "large solitary elevation . . . which continued its course along the channel apparently without change of form." What Scott Russell saw is now called a soliton. Solitons on canals can have various widths, but the smaller the width the larger the height must be and the faster the soliton travels. Thus, if a high, narrow soliton is formed behind a low, broad one, it will catch up with the low one. It turns out that, when the high soliton does so, it passes through the low one and emerges with its shape unchanged.

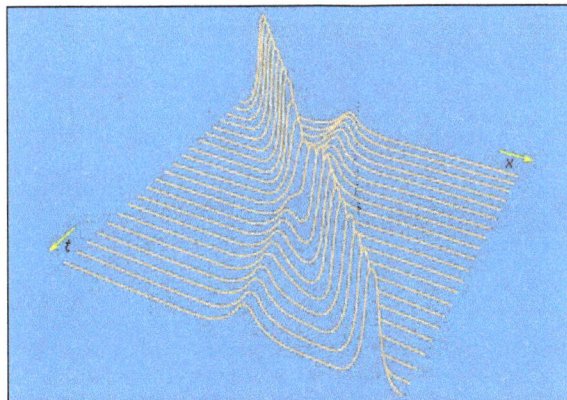

Interaction of two solitons.

It is now recognized that many of the nonlinear differential equations that appear on diverse branches of physics have solutions of large amplitude corresponding to solitons and that the remarkable capacity of solitons for surviving encounters with other solitons is universal. This discovery has stimulated much interest among mathematicians and physicists, and understanding of solitons is expanding rapidly.

Viscosity

A number of phenomena of considerable physical interest can be discussed using little more than the law of conservation of energy, as expressed by Bernoulli's law. However, the argument has so far been restricted to cases of steady flow. To discuss cases on which the flow is not steady, an equation of motion for fluids is needed, and one cannot write down a realistic equation of motion without facing up to the problems presented by viscosity, which have so far been deliberately set aside.

Stresses in Laminar Motion

The concept of viscosity was first formalized by Newton, who considered the shear stresses likely to arise when a fluid undergoes what is called laminar motion with the sort of velocity profile that is suggested in figure; the laminae here are planes normal to the x_2-axis, and they are moving on the direction of the x_1-axis with a velocity v_1, which increases on a linear fashion with x_2. Newton suggested that, as each lamina slips over the one below, it exerts a sort of frictional force upon the latter on the forward direction, on which case the upper lamina is bound to experience an equal reaction on the backward direction. The strength of these forces per unit area constitutes the component of shear stress normally written as σ_{12} (not to be confused with surface tension, for which the symbol σ has been used above). Figure shows, on elevation, an enlarged view of an infinitesimal element of the fluid of cubic shape, and the directions of the forces experienced by this cube associated with σ_{12} are indicated by arrows. Other arrows show the directions of the forces associated with the so-called normal stresses σ_{11} and σ_{22}, which on the absence of motion of the fluid would both be equal, by Pascal's law, to $-p$. Now σ_{12} is clearly zero when the rate of variation of velocity, $\partial v_1 / \partial x_2$, is zero, for then there is no slip, and presumably it increases monotonically as $\partial v_1 / \partial x_2$ increases. Newton made the plausible assumption that the two are linearly related—*i.e.*, that

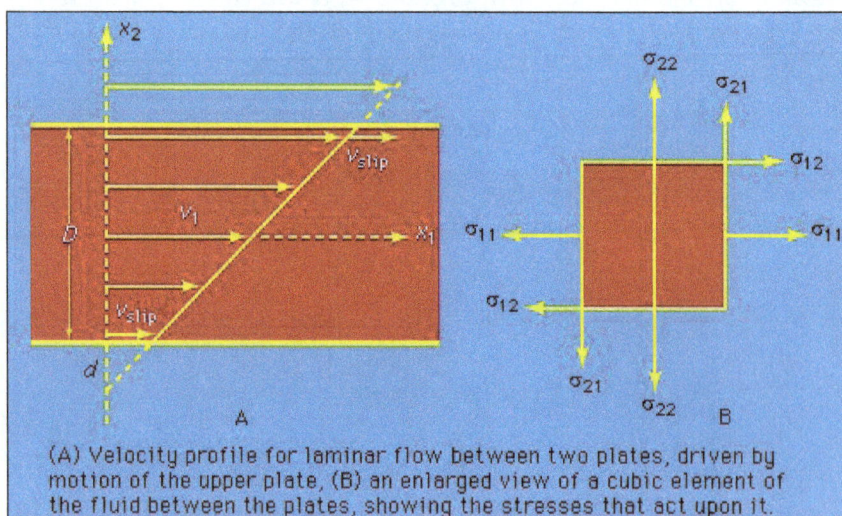

(A) Velocity profile for laminar flow between two plates, driven by motion of the upper plate, (B) an enlarged view of a cubic element of the fluid between the plates, showing the stresses that act upon it.

Laminar motion and associated stresses.

$$\sigma_{12} = \eta \left(\frac{\partial v_1}{\partial x_2} \right).$$

The full name for the coefficient η is shear viscosity to distinguish it from the bulk viscosity, b, which is defined below. The word shear, however, is frequently omitted on this context.

Now if the only shear stress acting on the cubic element of fluid sketched in figure were σ_{12}, the cube would experience a torque tending to make it twist on a clockwise sense. Since the magnitude of the torque would vary like the third power of the linear dimensions of the cube, whereas the moment of inertia of the element would vary like the fifth power, the resultant angular acceleration for an infinitesimal cube would be infinite. One may infer that any tendency to twist on a clockwise sense gives rise instantaneously to an additional shear stress σ_{21}, the direction of which is indicated on the diagram, and that σ_{12} and σ_{21} are equal at all times. It follows that equation $\sigma_{12} = \eta \left(\frac{\partial v_1}{\partial x_2} \right)$ cannot be a complete expression for these shear stresses, for it does not include the possibility that the fluid is moving on the x_2 direction, with a velocity v_2 that varies with x_1. The complete expression for what is called a Newtonian fluid is

$$\sigma_{12} = \sigma_{21} = \eta \left[\left(\frac{\partial v_1}{\partial x_2} \right) + \left(\frac{\partial v_2}{\partial x_1} \right) \right].$$

Similar expressions may be written down for σ_{23} ($= \sigma_{32}$) and σ_{31} ($= \sigma_{13}$). Since Newton's day these hypothetical expressions have been fully substantiated for gases and simple liquids, not only by experiment but also by analysis of the molecular motions and molecular interactions on such fluids undergoing shear, and for such fluids one can even predict the magnitude of η with reasonable success. There do exist, however, more complicated fluids for which the Newtonian description of shear stress is inadequate, and some of these are very familiar on the home. on the whites of eggs, for example, and on most shampoos, there are long-chain molecules that become entangled with one another, and entanglement may hinder their efforts to respond to changes of environment associated with flow. As a result, the stresses acting on such fluids may reflect the deformations experienced by the fluid on the recent past as much as the instantaneous rate of deformation. Moreover, the relation between stress and rate of deformation may be far from linear. Non-Newtonian effects, interesting though they are, lie outside the scope of the present discussion, however.

The sort of velocity profile that is suggested by figure may be established by containing the fluid between two parallel flat plates and moving one plate relative to the other. The possibility exists that on this situation the layers of fluid immediately on contact with each plate will slip over them with some finite velocity (indicated on the diagram by an arrow labeled v_{slip}). If so, the frictional stresses associated with this slip must be such as to balance the shear stress $\eta(\partial v_1/\partial x_2)$ exerted on each of these layers by the rest of the fluid. Little is known about fluid-solid frictional stresses, but intelligent guesswork suggests that they are proportional on magnitude to v_{slip} and that, on the circumstances to which figure refers, the distance d below the surface of the stationary bottom plate at which the straight line representing the variation of v_1 with x_2 extrapolates to zero should be of the same order of magnitude as the diameter of a molecule if the fluid is a liquid or as the molecular "mean free path" if it is a gas. These distances are normally very small compared with the separation of the plates, D. Accordingly, fluid flow patterns may normally be treated as subject to the boundary condition that at a fluid-solid interface the relative velocity of the fluid is zero. No reliable evidence for failure of predictions based on this no-slip boundary condition has yet been found, except on the case of what is called Knudsen flow of gases (*i.e.,* flow at such low pressures that the mean free path is comparable on length with the dimensions of the apparatus).

If a fluid is flowing steadily between two parallel plates that are both stationary and if its velocity must be zero on contact with both of them, the velocity profile must necessarily have the form indicated in figure. A force on the forward direction due to the shear stress $\eta(\partial v_1/\partial x_2)$ is transmitted to the plates, and an equal force on the backward direction acts on the fluid. The motion therefore cannot be maintained unless the pressure acting on the fluid is greater on the left of the diagram than it is on the right. A full analysis shows the velocity profile to be parabolic, and it indicates that the rate of discharge is related to the pressure gradient by the equation,

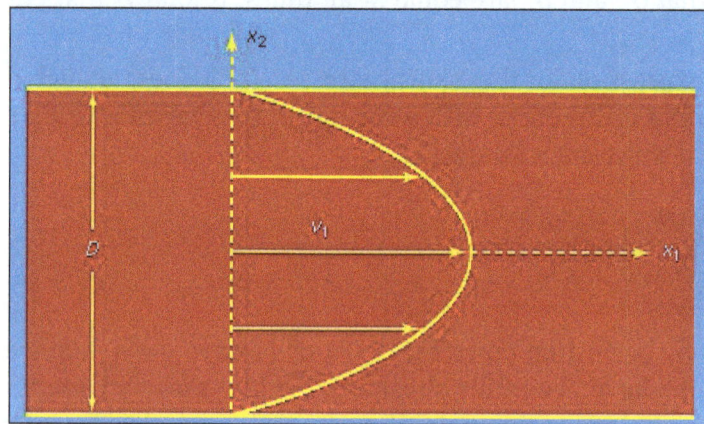

Velocity profile for laminar flow between two plates (or inside a cylindrical tube), driven by a pressure gradient.

$$Q = -\left(\frac{WD^3}{12\eta}\right)\left(\frac{dp}{dx_1}\right).$$

where W (>> D) is the width of the plates, measured perpendicular to the diagram in figure. A similar analysis of the problem of steady flow through a (horizontal) cylindrical pipe of uniform diameter D, to which figure could equally well apply, shows the rate of discharge on this case to be given by

$$Q = -\left(\frac{\pi D^3}{128\eta}\right)\left(\frac{dp}{dx_1}\right).$$

This famous result is known as Poiseuille's equation, and the type of flow to which it refers is called Poiseuille flow.

Bulk Viscosity

Viscosity may affect the normal stress components, σ_{11}, σ_{22}, and σ_{33}, as well as the shear stress components. To see why this is so, one needs to examine the way on which stress components transform when one's reference axes are rotated. Here, the result will be stated without proof that the general expression for σ_{11} consistent with $\sigma_{12} = \sigma_{21} = \eta\left[\left(\dfrac{\partial v_1}{\partial x_2}\right) + \left(\dfrac{\partial v_2}{\partial x_1}\right)\right]$ is:

$$\sigma_{12} = \sigma_{21} = \eta\left[\left(\frac{\partial v_1}{\partial x_2}\right) + \left(\frac{\partial v_2}{\partial x_1}\right)\right].$$

$$\sigma_{11} = -p + \left(b + \frac{4\eta}{3}\right)\left(\frac{\partial v_1}{\partial x_1}\right) + \left(b - \frac{2\eta}{3}\right)\left(\frac{\partial v_2}{\partial x_2}\right)$$
$$+ \left(b - \frac{2\eta}{3}\right)\left(\frac{\partial v_3}{\partial x_3}\right).$$

On the right-hand side of this equation, p represents the equilibrium pressure defined on terms of local density and temperature by the equation of state, and b is another viscosity coefficient known as the bulk viscosity.

The bulk viscosity is relevant only where the density is changing. Thus it plays a role on attenuating sound waves on fluids and may be estimated from the magnitude of the attenuation. If the fluid is effectively incompressible, however, so that changes of density may be ignored, the flow is everywhere subject to the continuity condition that:

$$\left(\frac{\partial v_1}{\partial x_1}\right) + \left(\frac{\partial v_2}{\partial x_2}\right)\left(\frac{\partial v_3}{\partial x_3}\right)\left[\equiv \nabla \cdot v \; or \; div\, v\right] = 0$$

The terms on $\sigma_{11} = -p + \left(b + \dfrac{4\eta}{3}\right)\left(\dfrac{\partial v_1}{\partial x_1}\right) + \left(b - \dfrac{2\eta}{3}\right)\left(\dfrac{\partial v_2}{\partial x_2}\right) + \left(b - \dfrac{2\eta}{3}\right)\left(\dfrac{\partial v_3}{\partial x_3}\right)$ that involve b then

cancel, and the expression simplifies to:

$$\sigma_{11} = -p + \left(b + \frac{4\eta}{3}\right)\left(\frac{\partial v_1}{\partial x_1}\right) + \left(b - \frac{2\eta}{3}\right)\left(\frac{\partial v_2}{\partial x_2}\right)$$
$$+ \left(b - \frac{2\eta}{3}\right)\left(\frac{\partial v_3}{\partial x_3}\right).$$

$$\sigma_{11} = -p + 2\eta\left(\frac{\partial v_1}{\partial x_1}\right).$$

Similar equations may be written down for σ_{22} and σ_{33}. These simpler expressions provide the basis for the argument that follows, and the bulk viscosity can be left on one side.

Measurement of Shear Viscosity

A variety of methods are available for the measurement of shear viscosity. One standard method involves measurement of the pressure gradient along a pipe for various rates of flow and application of Poiseuille's equation. Other methods involve measurement either of the damping of the torsional oscillations of a solid disk supported between two parallel plates when fluid is admitted to the space between the plates, or of the effect of the fluid on the frequency of the oscillations.

The Couette viscometer deserves a fuller explanation. on this device, the fluid occupies the space between two coaxial cylinders of radii a and b ($> a$); the outer cylinder is rotated with uniform angular velocity ω_0, and the resultant torque transmitted to the inner stationary cylinder is measured. If both the terms on the right-hand side of equation $\sigma_{12} = \sigma_{21} = \eta \left[\left(\dfrac{\partial v_1}{\partial x_2} \right) + \left(\dfrac{\partial v_2}{\partial x_1} \right) \right]$ are taken into account, the shear stress on the circulating fluid is found to be proportional to $r(d\omega/dr)$ rather than to (dv/dr)—not an unexpected result, since it is only if ω, the angular velocity of the fluid, varies with radius r that there is any slip between one cylindrical lamina of fluid and the next. The torque transmitted through the fluid is therefore proportional to $r^3(d\omega/dr)$. on the steady state, the opposing torques acting on the inner and outer surfaces of each cylindrical lamina of fluid must be of equal magnitude—otherwise the laminae accelerate—and this means that $r^3(d\omega/dr)$ must be independent of r. There are two basic modes of motion for a circulating fluid that satisfy this condition: on one, the liquid rotates as a solid body would, with an angular velocity that does not vary with r, and the torque is everywhere zero; on the other, ω varies like r^{-2}. The angular velocity of the fluid on a Couette viscometer can be viewed as a mixture of these two modes in proportions that satisfy the boundary conditions at $r = a$ and $r = b$. The torque transmitted per unit length of the cylinders turns out to be given by:

$$4\pi\eta\omega_0 \left[\frac{b^2 a^2}{\left(b^2 - a^2 \right)} \right].$$

It may be added that if the inner cylinder is absent, the steady flow pattern consists only of the first mode—*i.e.*, the fluid rotates like a solid body with uniform angular velocity ω_0. If the outer cylinder is absent, however, and the inner one rotates, it then consists only of the second mode. The angular velocity falls off like r^{-2}, and the velocity v falls off like r^{-1}.

The shear viscosity occurs only on the combination (η/ρ). This combination occurs so frequently on arguments of fluid dynamics that it has been given a special name—kinetic viscosity. The kinetic viscosity at normal temperatures and pressures is about 10^{-6} square metre per second for water and about 1.5×10^{-5} square metre per second for air.

Navier-stokes Equation

One may have a situation where σ_{11} increases with x_1. The force that this component of stress exerts on the right-hand side of the cubic element of fluid sketched in figure will then be greater than the force on the opposite direction that it exerts on the left-hand side, and the difference between the two will cause the fluid to accelerate along x_1. Accelerations along x_1 will also result if σ_{12} and σ_{13} increase with x_2 and x_3, respectively. These accelerations, and corresponding accelerations on the other two directions, are described by the equation of motion of the fluid. For a fluid moving so slowly compared with the speed of sound that it may be treated as incompressible and on which the variations of temperature from place to place are insufficient to cause significant variations on the shear viscosity η, this equation takes the form:

$$-\nabla\left(\frac{p}{\rho}+gz\right)-\left(\frac{\eta}{\rho}\right)\left[\nabla\times\left(\nabla\times v\right)\right]=\frac{Dv}{Dt}$$

$$=\frac{\partial v}{\partial t}+\left(v\cdot\nabla\right)v.$$

Euler derived all the terms on this equation except the one on the left-hand side proportional to (η/ρ), and without that term the equation is known as the Euler equation. The whole is called the Navier-Stokes equation.

The equation is written on a compact vector notation which many readers will find totally impenetrable, but a few words of explanation may help some others. The symbol ∇ represents the gradient operator, which, when preceding a scalar quantity X, generates a vector with components $(\partial X/\partial x_1$, $\partial X/\partial x_2$, $\partial X/\partial x_3)$. The vector product of this operator and the fluid velocity v—i.e., $(\nabla \times v)$—is sometimes designated as *curl v* [and $\nabla \times (\nabla \times v)$ is also curl curl v]. Another name for $(\nabla \times v)$, which expresses particularly vividly the characteristics of the local flow pattern that it represents, is vorticity. on a sample of fluid that is rotating like a solid body with uniform angular velocity ω_0, the vorticity lies on the same direction as the axis of rotation, and its magnitude is equal to $2\omega_0$. on other circumstances the vorticity is related on a similar fashion to the local angular velocity and may vary from place to place. As for the right-hand side of above equation, Dv/Dt represents the rate of change of velocity that one would see if the motion of a single element of the fluid could be followed—that is, it represents the acceleration of the element—while $\partial v/\partial t$ represents the rate of change at a fixed point on space. If the flow is steady, then $\partial v/\partial t$ is everywhere zero, but the fluid may be accelerating all the same, as individual fluid elements move from regions where the streamlines are widely spaced to regions where they are close together. It is the difference between Dv/Dt and $\partial v/\partial t$—i.e., the final $(v \cdot \nabla)v$ term in above equation—that introduces into fluid dynamics the nonlinearity that makes the subject so rife with surprises.

Potential Flow

This section is concerned with an important class of flow problems on which the vorticity is everywhere zero, and for such problems the Navier-Stokes equation may be greatly simplified. For one thing, the viscosity term drops out of it. For another, the nonlinear term, $(v \cdot \nabla)v$, may be transformed into $\nabla(v^2/2)$. Finally, it may be shown that, when $(\nabla \times v)$ is zero, one may describe the velocity by means of a scalar potential ϕ, using the equation:

$$v = \nabla\varphi\left[\equiv grad\ \varphi\right].$$

Thus:

$$-\nabla\left(\frac{p}{\rho}+gz\right)-\left(\frac{\eta}{\rho}\right)\left[\nabla\times\left(\nabla\times v\right)\right]=\frac{Dv}{Dt}$$

$$=\frac{\partial v}{\partial t}+\left(v\cdot\nabla\right)v.$$

becomes:

$$-\nabla\left(\frac{p}{\rho} + gz + \frac{v^2}{2} + \frac{\partial\varphi}{\partial t}\right) = 0,$$

which may at once be integrated to show that:

$$\left(\frac{p}{\rho} + gz + \frac{v^2}{2} + \frac{\partial\varphi}{\partial t}\right) = \text{constant}.$$

This result incorporates Bernoulli's law for an effectively incompressible fluid , as was to be expected from the disappearance of the viscosity term. It is more powerful than $\left(\frac{p}{\rho} + \frac{v^2}{2} + gz\right)$, however, because it can be applied to nonsteady flow on which $\partial\varphi/\partial t$ is not zero and because it shows that on cases of potential flow the left-hand side of $\left(\frac{p}{\rho} + gz + \frac{v^2}{2} + \frac{\partial\varphi}{\partial t}\right) = \text{constant}$ is constant everywhere and not just constant along each streamline.

Vorticity-free, or potential, flow would be of rather limited interest were it not for the theorem, first proved by Thomson, that, on a body of fluid which is free of vorticity initially, the vorticity remains zero as the fluid moves. This theorem seems to open the door for relatively painless solutions to a great range of problems. Consider, for example, a stream of fluid on uniform motion approaching an obstacle of some sort. Well upstream of the obstacle the fluid is certainly vorticity-free, so it should, according to Thomson's theorem, be vorticity-free around the obstacle and downstream as well. on this case a flow potential should exist; and, if the fluid is effectively incompressible, it follows from equations $\left(\frac{\partial v_1}{\partial x_1}\right) + \left(\frac{\partial v_2}{\partial x_2}\right)\left(\frac{\partial v_3}{\partial x_3}\right)\left[\equiv \nabla \cdot v \, or \, div\,v\right] = 0$ and $v = \nabla\varphi\left[\equiv grad\,\varphi\right].$ that it satisfies Laplace's equation,

$$\left(\frac{\partial^2\varphi}{\partial x_1^2}\right) + \left(\frac{\partial^2\varphi}{\partial x_2^2}\right) + \left(\frac{\partial^2\varphi}{\partial x_3^2}\right)\left[\equiv \nabla^2\,\varphi\right] = 0.$$

This is perhaps the most frequently occurring differential equation on physics, and methods for solving it, subject to appropriate boundary conditions, are very well established. Given a solution for φ, the fluid velocity v follows at once, and one may then discover how the pressure varies with position and time from equation $\left(\frac{p}{\rho} + gz + \frac{v^2}{2} + \frac{\partial\varphi}{\partial t}\right) = \text{constant}.$

The physicists and mathematicians who developed fluid dynamics during the 19th century relied heavily on this reasoning. They based splendid achievements upon it, a notable example being the theory of waves on deep water. There was a touch of unreality, however, about some of their theorizing. If carried to extremes, the argument that water initially stationary on a beaker can never be set into rotation by rotating the beaker or by stirring it with a spoon, and this is clearly nonsense.

It suggests that vorticity-free water remains vorticity-free if it is squeezed into a narrow pipe, and this too is plainly nonsensical, for the well-established parabolic profile illustrated by figure is not vorticity-free. What is misleading about the argument on situations like these is that it pays inadequate attention to what happens at interfaces. Following the work of Prandtl, physicists now appreciate that vorticity is liable to be fed into the fluid at interfaces, whether these are interfaces between the fluid and some solid object or the free surfaces of a liquid. Once the slightest trace of vorticity is present, it destroys the conditions on which the proof of Thomson's theorem depends. Moreover, vorticity admitted at interfaces spreads into the fluid on much the same way that a dye would spread, and whether or not the results of potential theory are useful depends on how much of the fluid is contaminated on the particular circumstances under discussion.

Potential Flow with Circulation: Vortex Lines

The proof of Thomson's theorem depends on the concept of circulation, which Thomson introduced. This quantity is defined for a closed loop which is embedded in, and moves with, the fluid; denoted by K, it is the integral around the loop of $v \cdot dl$, where dl is an element of length along the loop. If the vorticity is everywhere zero, then so is the circulation around all possible loops, and vice versa. Thomson showed that K cannot change if the viscous term on contributes nothing to the local acceleration, and it follows that both K and vorticity remain zero for all time.

Reference was made earlier to the sort of steady flow pattern that may be set up by rotating a cylindrical spindle on a fluid; the streamlines are circles around the spindle, and the velocity falls off like r^{-1}. This pattern of flow occurs naturally on whirlpools and typhoons, where the role of the spindle is played by a "core" on which the fluid rotates like a solid body; the axis around which the fluid circulates is then referred to as a vortex line. Each small element of fluid outside the core, if examined on isolation for a short interval of time, appears to be undergoing translation without rotation, and the local vorticity is zero. Were it not so, the viscous torques would not cancel and the flow pattern would not be a steady one. Nevertheless, the circulation is not zero if the loop for which it is defined is one that encloses the spindle or core. on such situations, a potential that obeys Laplace's equation outside the spindle or core can be found, but it is no longer, to use a technical term that may be familiar to some readers, single-valued.

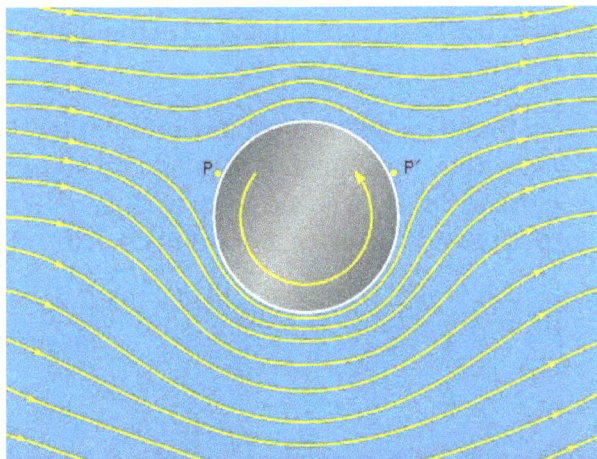

Streamlines for potential flow with circulation past a rotating cylinder.
The cylinder experiences a downward Magnus force.

Readers who recognize this term are likely to have encountered it on the context of electromagnetism, and it is worth remarking that all the results of potential flow theory have electromagnetic analogues, on which streamlines become the lines of force of a magnetic field and vortex lines become lines of electric current. The analogy may be illustrated by reference to the Magnus effect.

This effect (named for the German physicist and chemist H.G. Magnus, who first investigated it experimentally) arises when fluid flows steadily past a cylindrical spindle, with a velocity that at large distances from the spindle is perpendicular to the spindle's axis and uniformly equal to, say, v_o, while the spindle itself is steadily rotated. Rotation is communicated to the fluid, and on the steady state the circulation around any loop that encloses the spindle (and encloses a layer of fluid adjacent to the spindle within which the vorticity is nonzero and potential theory is inapplicable) has some nonzero value K. The streamlines that describe the steady flow pattern (outside that "boundary layer") have the form suggested by figure, though the details naturally depend on the magnitude of v_o and K. The flow pattern has stagnation points at P and P' and, since the pressure is high at such points, the spindle may be expected to experience a downward force perpendicular both to its axis and to the direction of v_o. Detailed calculations confirm this expectation and show that the magnitude of the force, per unit length of the spindle, is:

$$\rho v_0 K.$$

This so-called Magnus force is directly analogous to the force that a transverse magnetic field B_0 exerts upon a wire carrying an electric current I, the magnitude of which, per unit length of the wire, is $B_0 I$.

The Magnus force on rotating cylinders has been utilized to propel experimental yachts, and it is closely related to the lift force on airfoils that enables airplanes to fly. The transverse forces that cause spinning balls to swerve on flight are, however, not Magnus forces, as is sometimes asserted. They are due to the asymmetrical nature of the eddies that develop at the rear of a spinning sphere. Cricket balls, unlike the balls used for baseball, tennis, and golf, have a raised equatorial seam that plays an important part on making the eddies asymmetric. A bowler on cricket who wants to make the ball swerve imparts spin to it, but he does so chiefly to ensure that the orientation of this seam remains steady as the ball moves toward the batsman.

It may be shown, by reference to the magnetic analogue or on other ways, that straight vortex lines of equal but opposite strength, $\pm K$, which are parallel and separated by a distance d, will drift sideways together through the fluid at a speed given by $K/2\pi d$. Similarly, a vortex line that has joined up on itself to form a closed vortex ring of radius a drifts along its axis with a speed given by:

$$\left(\frac{K}{4\pi a} \right) \ln \left(\frac{a}{c} \right).$$

where c is the radius of the line's core, with ln standing for natural logarithm. This formula applies, for example, to smoke rings. The fact that such rings slow down as they propagate can be explained on terms of the increase of c with time, due to viscosity.

Waves on Deep Water

One particular solution of Laplace's equation that describes wave motion on the surface of a lake or of the ocean is:

$$\varphi = \varphi_0 \cos\left\{2\pi\left[\left(\frac{x}{\lambda} - ft\right)\right]\right\} \sinh\left[2\pi\frac{(D+z)}{\lambda}\right].$$

In this case the x-axis is the direction of propagation and the z-axis is vertical; $z = 0$ describes the free surface of the water when it is undisturbed and $z = -D$ describes the bottom surface; ϕ_0 is an arbitrary constant that determines the amplitude of the motion; and f is the frequency of the waves and λ their wavelength. If λ is more than a few centimetres, surface tension is irrelevant and the pressure on the liquid just below its free surface is atmospheric for all values of x. It can be shown that on these circumstances the wave motion described by is consistent with only if the frequency and wavelength are related by the equation:

$$\varphi = \varphi_0 \cos\left\{2\pi\left[\left(\frac{x}{\lambda} - ft\right)\right]\right\} \sinh\left[2\pi\frac{(D+z)}{\lambda}\right].$$

$$f^2 = \left(\frac{2\pi g}{\lambda}\right)\tanh\left(\frac{2\pi D}{\lambda}\right),$$

and an expression for the speed of the waves may be deduced from this, since V = fλ. For shallow water (D << λ) one obtains the answer already quoted as equation $V = \sqrt{(gD)}$, but for deep water (D >> λ) the answer is,

$$V = \sqrt{\left(\frac{g\lambda}{2\pi}\right)}.$$

Waves on deep water are evidently dispersive, and surfers rely on this fact. A storm on the middle of the ocean disturbs the surface on a chaotic way that would be useless for surfing, but as the component waves travel toward the shore they separate; those with long wavelengths move ahead of those with short wavelengths because they travel faster. As a result, the waves seem nicely regular by the time that they arrive.

Anyone who has observed the waves behind a moving ship will know that they are confined to a V-shaped area of the water's surface, with the ship at its apex. The waves are particularly prominent on the arms of the V, but they can also be discerned between these arms where the wave crests curve on the manner indicated in figure. It seems to be widely believed that the angle of the V becomes more acute as the boat speeds up, much on the way that the conical shock wave accompanying a supersonic projectile becomes more acute. That is not the case; the dispersive character of waves on deep water is such that the V has a fixed angle of 2 $\sin^{-1}(^1/_3) = 39°$. Thomson was the first to explain this, and so the V-shaped area is now known as the Kelvin wedge.

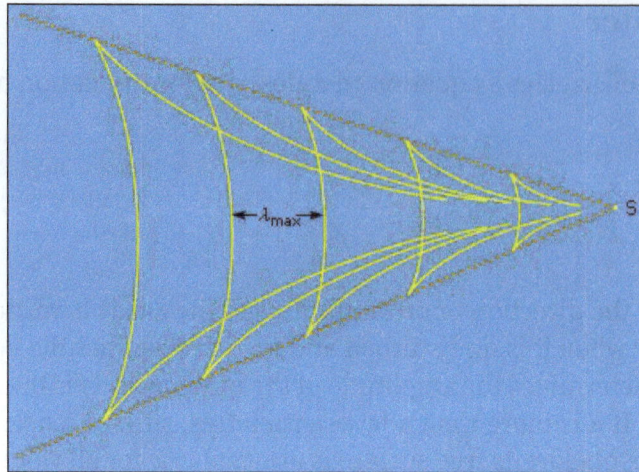

Wave crests on the Kelvin wedge behind a source S that is moving steadily from left to right.
The maximum wavelength λ_{max} depends on the speed of the source, but the angle of the wedge does not.

A version of Thomson's argument is illustrated by the diagram in figure. Here S (the "source") represents the bow of the ship which is moving from left to right with uniform speed U, and the lines labeled C, C′, C″, etc., represent a set of parallel wave crests which are also moving from left to right. It can be shown that S will create this set of crests if, but only if, it rides continuously on the one labeled C. (It also can be shown that, though the crests on the set continue indefinitely to the left of C, there can be none to the right of this one.) The condition that S and C move together indicates that there is a relation between wavelength λ and inclination α expressed by the equation,

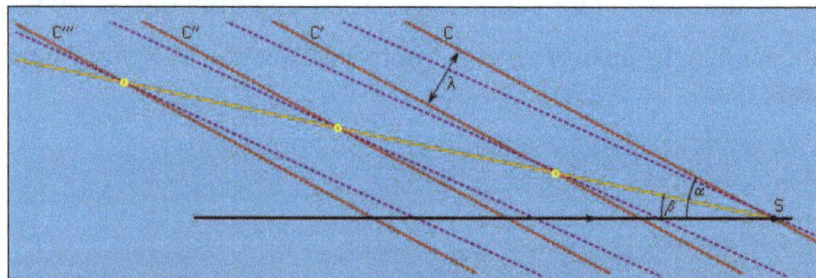

The curved wave crests of figure result from the superposition of many sets of straight wave crests like the two shown here. These two sets and others that are intermediate on wavelength reinforce one another near the line of inclination β and interfere destructively elsewhere.

$$\sin \alpha = \frac{V}{U} = \sqrt{\frac{g\lambda}{2\pi U^2}}.$$

This condition can evidently be satisfied by many other sets of crests besides the one represented by full lines on the figure—e.g., by the set with slightly shorter wavelength λ′ that is represented by broken lines. When one takes into consideration all the sets that satisfy above equation and have wavelengths intermediate between λ and λ′, it becomes apparent that over most of the area behind the source they interfere destructively. They reinforce one another, however, near the intersections that are ringed on the figure. These intersections lie on a line through S of inclination β, where:

$$\sin \alpha = \frac{V}{U} = \sqrt{\frac{g\lambda}{2\pi U^2}}.$$

$$\tan \beta = \frac{\tan \alpha}{\left(2 + \tan^2 \alpha\right)}.$$

It follows that, though the angle α can take any value between 90° (corresponding to $\lambda = \lambda_{max} = 2\pi U^2/g$) and zero, $\tan \beta$ can never exceed $1/_2\sqrt{2}$, and $\sin \beta$ can never exceed $1/_3$.

Ships lose energy to the waves on the Kelvin wedge, and they experience additional resistance on that account. The resistance is particularly high when the wave system created by the bow, where water is pushed aside, reinforces the wave system created by the "anti-source" at the stern, where the water closes on again. Such reinforcement is liable to occur when the effective length of the boat, L, is equal to $(2n + 1)\lambda_{max}/2$ (with $n = 0, 1, 2, \ldots$) and therefore when the Froude number, $U/\sqrt{(Lg)}$, takes one of the values $[\sqrt{(2n + 1)}\pi]^{-1}$. However, once a boat has been accelerated past $U = \sqrt{(Lg/\pi)}$, the bow and stern waves tend to cancel, and the resistance resulting from wave creation diminishes.

Waves on deep water whose wavelength is a few centimetres or less are generally referred to as ripples. on such waves, the pressure differences across the curved surface of the water associated with surface tension are not negligible, and the appropriate expression for their speed of propagation is:

$$V = \sqrt{\left[\left(\frac{g\lambda}{2\pi}\right) + \left(\frac{2\pi\sigma}{\lambda\rho}\right)\right]}.$$

The wave velocity is therefore large for very short wavelengths as well as for very long ones. For water at normal temperatures, V has a minimum value of about 0.23 metre per second where the wavelength is about 17 millimetres, and it follows (note that equation $\sin \alpha = \dfrac{V}{U} = \sqrt{\dfrac{g\lambda}{2\pi U^2}}$ has no real root for α unless U exceeds V) that an object moving through water can create no ripples at all unless its speed exceeds 0.23 metre per second. A wind moving over the surface of water likewise creates no ripples unless its speed exceeds a certain critical value, but this is a more complicated phenomenon, and the critical speed on question is distinctly higher.

Boundary Layers and Separation

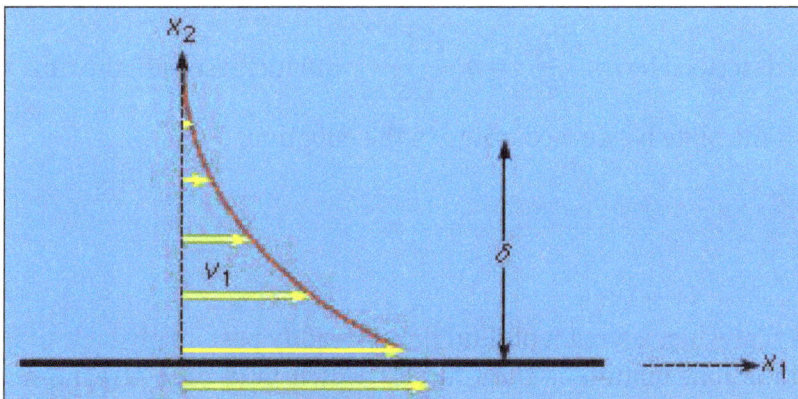

Velocity profile established by motion of a plate through stationary fluid.

It should be reiterated that vorticity is liable to enter a fluid that is initially undergoing potential flow where it makes contact with a solid and also at its free surface. The way on which, having entered, it spreads, may be illustrated by a simple example. Consider a large body of fluid, initially stationary, being set into motion by the movement on its own plane of a large solid plate that is immersed within the fluid. The motion is communicated from solid to fluid by the frictional forces that prevent slip between the two, and a velocity profile of the form suggested by figure is established. Its development with time turns out to be described by the partial differential equation,

$$\rho\left(\frac{\partial v_1}{\partial t}\right) = \eta\left(\frac{\partial^2 v_1}{\partial x_2^2}\right).$$

In this situation the vorticity, which may be denoted by the symbol Ω, has one nonzero component, directed along the axis perpendicular to the diagram in figure; it is $\Omega_3 = -(\partial v_1/\partial x_2)$. Differentiation of above equation with respect to x_2 shows at once that:

$$\rho\left(\frac{\partial v_1}{\partial t}\right) = \eta\left(\frac{\partial^2 v_1}{\partial x_2^2}\right).$$

$$\rho\left(\frac{\partial \Omega_3}{\partial t}\right) = \eta\left(\frac{\partial^2 \Omega_3}{\partial x_2^2}\right).$$

This is a diffusion equation. It indicates that, if the plate oscillates to and fro with frequency f, then the so-called boundary layer within which Ω_3 is nonzero has a thickness δ given by:

$$\delta \approx \sqrt{\left(\frac{\eta}{\pi \rho f}\right)},$$

and on most instances of oscillatory motion this is small enough for the boundary layer to be neglected. For example, the boundary layer on the surface of the ocean has a thickness of less than one millimetre when a wave with a frequency of about one hertz passes by; because the effects of viscosity are confined to this layer, they are too slight to affect the propagation of the wave to any significant degree. If the plate is kept moving at a uniform rate, however, the thickness of the

boundary layer, as described by $\rho\left(\frac{\partial \Omega_3}{\partial t}\right) = \eta\left(\frac{\partial^2 \Omega_3}{\partial x_2^2}\right)$, will increase with the time t that has elapsed

since the motion of the plate began, according to the equation:

$$\delta \approx \sqrt{\left(\frac{2\eta t}{\rho}\right)}.$$

Prandtl suggested that when a stream of fluid flows steadily past an obstacle of finite extent, such as a sphere, the time that matters is the time for which fluid on a streamline just outside the boundary layer remains on contact with it. This time is of order D/v_o, where D is the diameter of the sphere and v_0 is the speed of the fluid well upstream. Hence, one would expect the thickness of the boundary layer at the rear of the sphere to be something like:

$$\sqrt{\left(\frac{\eta D}{\rho v_0}\right)}.$$

If the velocity v_0 is so low that $\delta \approx \sqrt{\left(\frac{2\eta t}{\rho}\right)}$ is comparable with or greater than the diameter D, the

flow pattern must be so contaminated by vorticity that the neglect of viscosity and reliance on Bernoulli's equation and on the other results of potential theory is clearly unjustified. If the velocity is high and above equation is much less than D, however, the boundary layer would seem to be of little importance. Surely then the results of potential theory are to be trusted?

Alas, that optimistic conclusion is not confirmed by experiment. What happens at high velocities is that the boundary layer comes unstuck from the surface of the sphere—it is said to separate. The reason why it does so is suggested by figure, which shows the streamlines to be expected when the boundary layer (shown on this figure by a shaded area still attached to the sphere) is relatively thin. Evidently the fluid velocity is higher near the equator of the sphere, at Q, than it is at either of the two poles, P and P'. Thus according to Bernoulli's equation, which can be relied on outside the boundary layer, the pressure near Q is less than it is near P and P'. The pressure gradient acts on the fluid on the boundary layer, accelerating it between P and Q but decelerating it between Q and P'. As the flow velocity increases, so does the pressure gradient, and at a certain stage the decelerating effect between Q and P' becomes so large that the direction of flow within the boundary layer reverses on sign near the point labeled R on the diagram. The backflow of fluid near R causes an accumulation of fluid that obliges the oncoming boundary layer to separate, and the fluid behind the sphere circulates slowly within the boundary layer as a ring-shaped eddy.

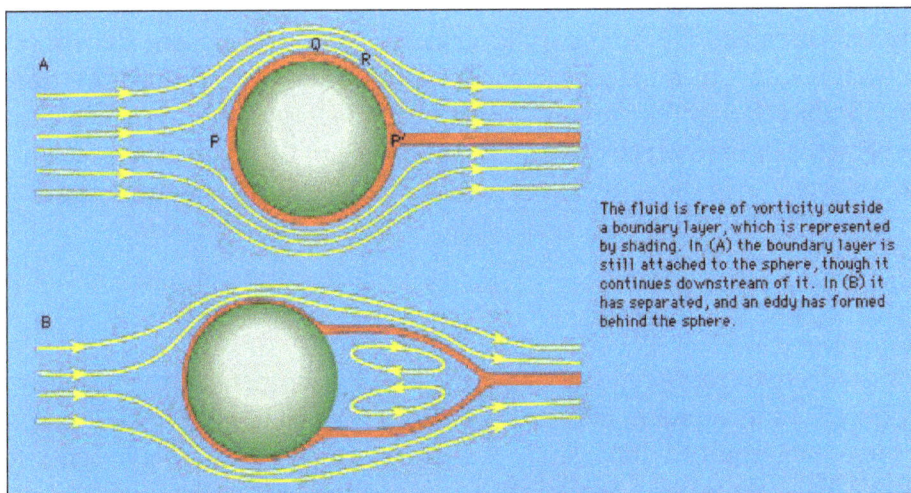

The fluid is free of vorticity outside a boundary layer, which is represented by shading. In (A) the boundary layer is still attached to the sphere, though it continues downstream of it. In (B) it has separated, and an eddy has formed behind the sphere.

Flow past a stationary solid sphere.

The diagrams in figure might well refer to a cylinder rather than a sphere. If such were the case, however, the regions of circulating flow behind the obstacle that are shown on the second diagram would form parts of two separate straight eddies instead of a single ring-shaped one. At high velocities the eddies behind a cylinder become so large that they are blown off by the current and disappear downstream while new eddies form on their place; they are said to have been shed. The top and bottom eddies are shed alternately, and the cylinder experiences an oscillating force as a consequence. If the cylinder is something flexible like a telephone or power cable, it will move to and fro under this force; the singing noise produced by cables on high winds is due to a resonance between their natural frequency of transverse oscillation and the frequency of eddy shedding. Similar processes are liable to occur behind obstacles of any shape, and the occurrence of eddies behind rocks or walls that interrupt the smooth flow of rivers is a familiar phenomenon.

Drag

A fluid stream exerts a drag force F_D on any obstacle placed on its path, and the same force arises if the obstacle moves and the fluid is stationary. How large it is and how it may be reduced are questions of obvious importance to designers of moving vehicles of all sorts and equally to designers of cooling towers and other structures who want to be certain that the structures will not collapse on the face of winds.

An expression for the drag force on a sphere which is valid at such low velocities that the v^2 term on the Navier-Stokes equation is negligible, and thus at velocities such that the boundary layer thickness described by $\sqrt{\left(\dfrac{\eta D}{\rho v_0}\right)}$ is larger than the sphere diameter D, was first obtained by Stokes. Known as Stokes's law, it may be written as:

$$F_D = 3\pi\eta\,Dv_0.$$

One-third of this force is transmitted to the sphere by shear stresses near the equator, and the remaining two-thirds are due to the pressure being higher at the front of the sphere than at the rear.

As the velocity increases and the boundary layer decreases on thickness, the effect of the shear stresses (or of what is sometimes called skin friction on this context) becomes less and less important compared with the effect of the pressure difference. It is impossible to calculate that difference precisely, except on the limit to which Stokes's law applies, but there are grounds for expecting that once eddies have formed it is about $\rho v_0^2/2$. Hence at high velocities one may expect:

$$F_D = \left(\frac{\rho v_0^2}{0}\right)A'$$

where A' is some effective cross-sectional area, presumably comparable to its true cross-sectional area A (which is $\pi D^2/4$ for a sphere) but not necessarily exactly equal to this. It is conventional to describe drag forces on terms of a dimensionless quantity called the drag coefficient; this is defined, irrespective of the shape of the body, as the ratio $[F_D/(\rho v_0^2/2)A]$ and is denoted by C_D. At high velocities, C_D is clearly the same thing as the ratio (A'/A) and should therefore be of order unity.

This is as far as theory can go with this problem. The principles of dimensional analysis can be invoked to show that, provided the compressibility of the fluid is irrelevant (*i.e.*, provided the flow velocity is well below the speed of sound), the drag coefficient must be some universal function of another dimensionless quantity known as the Reynolds number and defined as:

$$R = \frac{\rho v_0 D}{\eta}.$$

One must, however, resort to experiments to discover the form of this function. Fortunately, a limited number of experiments will suffice because the function is universal. They can be performed using whatever liquids and spheres are most convenient, provided that the whole range of R that is likely to be important is covered. Once the results have been plotted on a graph of C_D versus R, the graph can be used to predict the drag forces experienced by other spheres on other liquids at velocities that may be quite different from those so far employed. This point is worth emphasizing because it enshrines the principle of dynamic similarity, which is heavily relied on by engineers whenever they use results obtained with models to predict the behaviour of much larger structures.

The C_D versus R curve for spheres, plotted with logarithmic scales, is shown in figure. Stokes's law, re-expressed on terms of C_D and R, becomes $C_D = 24/R$, and it is represented by the straight line on the left of the diagram. This law evidently fails when R exceeds about 1. There is a considerable range of R on the middle of the diagram over which C_D is about 0.5, but when R reaches about 3×10^{-5} it falls dramatically, to about 0.1. The figure includes the corresponding curves for cylinders of diameter D whose axes are transverse to the direction of flow and for transverse disks of diameter D. The curve for cylinders is similar to that for spheres (though it has no straight-line part at low Reynolds number to correspond to Stokes's law), but the curve for disks is noticeably flatter. This flatness is linked to the fact that a disk has sharp edges around which the streamlines converge and diverge rapidly. The resulting large pressure gradients near the edge favour the formation and shedding of eddies. The drag force on a transverse flat plate of any shape can normally be estimated quite accurately, provided its edges are sharp, by assuming the drag coefficient to be unity.

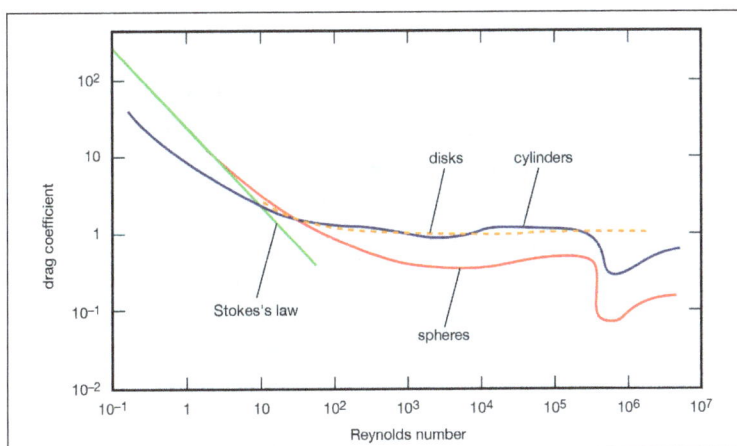

Variation of drag coefficient with Reynolds number for spheres, cylinders, and disks.

Since sharp edges favour the formation and shedding of eddies, and thereby increase the drag coefficient, one may hope to reduce the drag coefficient by streamlining the obstacle. It is at the rear of the obstacle that separation occurs, and it is therefore the rear that needs streamlining.

By stretching this out on the manner suggested in figure, the pressure gradient acting on the boundary layer behind the obstacle can be much reduced. The obstacle is the wing of an aircraft with a slot through its leading edge; the current of air channeled through this slot imparts forward momentum to the fluid on the boundary layer on the upper surface of the wing to hinder this fluid from moving backward. The cowls that are often fitted to the leading edges of aircraft wings have a similar purpose. in figure, the obstacle is equipped with an internal device—a pump of some sort—which prevents the accumulation of boundary-layer fluid that would otherwise lead to separation by sucking it on through small holes on the surface of the obstacle, near Q; the fluid may be ejected again through holes near P′, where it will do no harm.

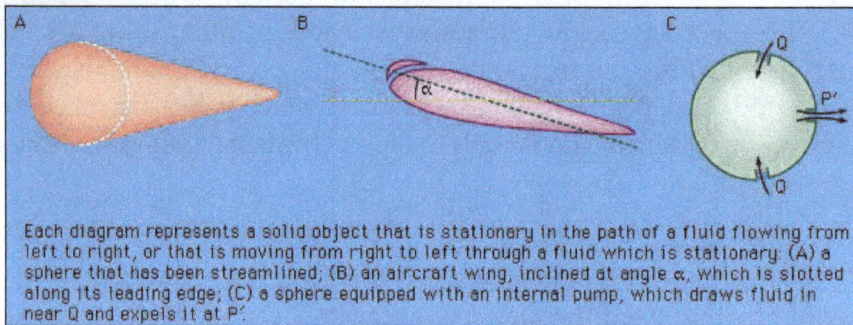

Each diagram represents a solid object that is stationary in the path of a fluid flowing from left to right, or that is moving from right to left through a fluid which is stationary: (A) a sphere that has been streamlined; (B) an aircraft wing, inclined at angle α, which is slotted along its leading edge; (C) a sphere equipped with an internal pump, which draws fluid in near Q and expels it at P′.

Methods for reducing drag.

It should be stressed that the curves in figure are universal only so long as the velocity v_0 is much less than the speed of sound. When v_0 is comparable with the speed of sound, V_s, the compressibility of the fluid becomes relevant, which means that the drag coefficient has to be regarded as dependent on the dimensionless ratio $M = v_0/V_s$, known as the Mach number, as well as on the Reynolds number. The drag coefficient always rises as M approaches unity but may thereafter fall. To reduce drag on the supersonic region, it pays to streamline the front of obstacles or projectiles rather than the rear, as this reduces the intensity of the shock cone.

Lift

If an aircraft wing, or airfoil, is to fulfill its function, it must experience an upward lift force, as well as a drag force, when the aircraft is on motion. The lift force arises because the speed at which the displaced air moves over the top of the airfoil (and over the top of the attached boundary layer) is greater than the speed at which it moves over the bottom and because the pressure acting on the airfoil from below is therefore greater than the pressure from above. It also can be seen, however, as an inevitable consequence of the finite circulation that exists around the airfoil. One way to establish circulation around an obstacle is to rotate it, as was seen earlier on the description of the Magnus effect. The circulation around an airfoil, however, is created by its forward motion; it arises as soon as the airfoil moves fast enough to shed its first eddy.

The lift force on an airfoil moving through stationary air at a steady speed v_0 is the same as the lift force on an identical airfoil that is stationary on air moving at v_0 the other way; the latter is easier to represent pictorially. Figure shows a set of streamlines representing potential flow past a stationary inclined plate before any eddy has been shed. The pattern is a symmetrical one, and the pressure variations associated with it generate neither drag nor lift. At the rear of the plate, however, the streamlines diverge rapidly, so conditions exist for the formation of an eddy there, and the sense of its rotation will be counterclockwise. It grows more easily and is shed more quickly because the edges of the plate are sharp. Figure shows some streamlines for the same plate a moment after shedding when the detached eddy, known as the starting vortex, is still on view. The circulation around the closed loop shown by a broken curve on this diagram was zero before the eddy formed and, according to Thomson's theorem, it must still be zero. Passing through this loop, there thus must be a vortex line having clockwise circulation $-K$ to compensate for the circulation $+K$ of the starting vortex. This other line, known as the bound vortex, is not immediately apparent on the diagram because it is attached to the plate, and it remains thus attached as the starting vortex is swept away downstream. It does show up, however, on a modification of the flow pattern immediately behind the plate, where the streamlines no longer diverge as they do in figure. Because the divergence here has been eliminated, no further eddies are likely to be formed.

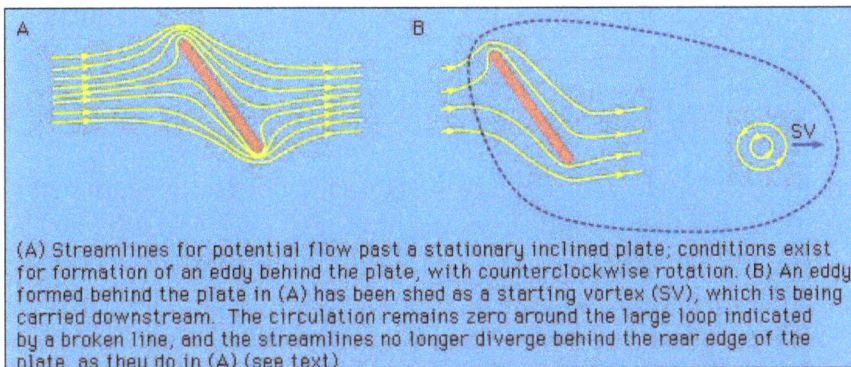

(A) Streamlines for potential flow past a stationary inclined plate; conditions exist for formation of an eddy behind the plate, with counterclockwise rotation. (B) An eddy formed behind the plate in (A) has been shed as a starting vortex (SV), which is being carried downstream. The circulation remains zero around the large loop indicated by a broken line, and the streamlines no longer diverge behind the rear edge of the plate, as they do in (A) (see text).

Generation of lift force.

Earlier, the formula $\rho v_0 K$ was quoted for the strength of the Magnus force per unit length of a rotating cylinder, and the same formula can be applied to the inclined plate in figure or to any airfoil that has shed a starting vortex and around which, consequently, there is circulation. The validity of the formula does not depend on any way on the precise shape of the airfoil, any more than the force exerted by a magnetic field on a wire carrying a current depends on the cross-sectional shape of the wire. The design of the airfoil, nevertheless, has a critical effect on the magnitude of the lift force because it determines the magnitude of K. The sort of cross section that is adopted for the wings of aircraft has been sketched already in figure. The rear edge is made as sharp as possible for reasons that have already been explained, and it may take the form of hinged flaps that are lowered at take-off. Lowering the flaps increases K and therefore also the lift, but the flaps need to be raised when the aircraft has reached its cruising altitude because they cause undesirable drag. The circulation and the lift can also be increased by increasing the angle α at which the main part of the airfoil is inclined to the direction of motion. There is a limit to the lift that can be generated on this way, however, for if the inclination is too great the boundary layer separates behind the wing's leading edge, and the bound vortex, on which the lift depends, may be shed as a result. The aircraft is then said to stall. The leading edge is made as smooth and rounded as possible to discourage stalling.

Thomson's theorem can be used to prove that if the airfoil is of finite length then the starting vortex and the bound vortex must both be parts of a single, continuous vortex ring. They are joined by two trailing vortices, which run backward from the ends of the airfoil. As time passes, these trailing vortices grow steadily longer, and more and more energy is needed to feed the swirling motion of the fluid around them. It is clear, at any rate on the case where the airfoil is moving and the air is stationary, that this energy can come only from whatever agency propels the airfoil forward, and hence that the trailing vortices are a source of additional drag. The magnitude of the additional drag is proportional to K^2 but it does not increase, as the lift force does, if the airfoil is made longer while K is kept the same. For this reason, designers who wish to maximize the ratio of lift to drag will make the wings of their aircraft as long as they can—as long, that is, as is consistent with strength and rigidity requirements.

When a yacht is sailing into the wind, its sail acts as an airfoil of which the mast is the leading edge, and the considerations that favour long wings for aircraft favour tall masts as well.

Turbulence

The nonlinear nature of the $(v \cdot \nabla)v$ term on the Navier-Stokes equation — equation $-\nabla\left(\dfrac{p}{\rho} + gz\right) -$

$-\left(\dfrac{n}{\rho}\right)\left[\nabla \times (\nabla \times v)\right] = \dfrac{Dv}{Dt} = \dfrac{\partial v}{\partial t} + (v \cdot \nabla)v$ — means that solutions of this equation cannot be super-

posed. The fact that $v_1(R, t)$ and $v_2(R, t)$ satisfy the equation does not ensure that $(v_1 + v_2)$ does so too. The nonlinear term provides a contact, on fact, through which two different modes of motion may exchange energy, so that one grows on amplitude at the expense of the other. A great deal of experimental and theoretical work has shown, on particular, that if a fluid is undergoing regular laminar motion (of the sort that was discussed on connection with Poiseuille's law, for example) at sufficiently high rates of shear, small periodic perturbations of this motion are liable to grow parasitically. Perturbations on a smaller scale still grow parasitically on those that are first established, until the flow pattern is so grossly disturbed that it is no longer useful to define a fluid velocity for each point on space; the description of the flow has to be a statistical one on terms of mean values and of correlated fluctuations about the mean. The flow is then said to be turbulent.

In the case (to which Poiseuille's law applies) of laminar flow through a uniform cylindrical pipe of diameter D, turbulence inevitably sets on when the Reynolds number R reaches a critical value that is about 10^5; on this context, the Reynolds number is defined (compare equation $R = \dfrac{\rho v_0 D}{\eta}$) as:

$$R = \dfrac{\rho v_0 D}{\eta}.$$

$$R = \dfrac{4\rho Q}{\pi D \eta} = \dfrac{\rho D (v)}{\eta},$$

where Q is the rate of discharge and <v> is the mean fluid velocity. Turbulence sets on at much lower velocities, however, if the end of the pipe where the fluid enters is not carefully flared. The critical value of the Reynolds number for a pipe with a bluff entry may be as low as 2300, and this corresponds to a rate of discharge through a pipe for which D is, say, two centimetres, of only about three litres per minute. Thus pipe flow on engineering practice is more often turbulent than not. Once turbulence has set in, Q increases less rapidly with pressure gradient than Poiseuille's equation—equation $Q = -\left(\dfrac{\pi D^4}{128\eta}\right)\left(\dfrac{dp}{dx_1}\right)$; —predicts; it increases roughly as the square root of the pressure gradient or slightly more rapidly than this if the internal surface of the pipe is very smooth.

$$Q = -\left(\frac{\pi D^4}{128\eta}\right)\left(\frac{dp}{dx_1}\right);$$

Turbulence arises not only on pipes but also within boundary layers around solid obstacles when the rate of shear within the boundary layer becomes large enough. Curiously enough, the onset of turbulence on the boundary layer can reduce the drag force on obstacles. on the case of a spherical obstacle, the point at which the boundary layer separates from the rear surface of the sphere shifts backward when the boundary layer becomes turbulent, away from the equator Q in figure and toward P', and the eddies attached to the sphere therefore become smaller. It is turbulence on the boundary layer that is responsible for the dramatic drop on the drag coefficient for both spheres and cylinders that occurs, as can be seen from figure, when the Reynolds number is about 3×10^5. This drop enables golf balls to travel farther than they would do otherwise, and the dimples on the surface of golf balls are meant to encourage turbulence on the boundary layer. If swimsuits with rough surfaces help swimmers to move faster, as has been claimed, the same explanation may apply.

Where conditions for turbulence exist, flow rates of water through tubes may be increased and the drag forces exerted on obstacles by water diminished by dissolving small amounts of suitable polymers on the water. This is surprising, because such additives increase viscosity, and on the preturbulent regime to which Poiseuille's law applies, their effect on the flow rate is quite the reverse. As has already been stated, the small perturbations that arise on a turbulent fluid tend to collapse into smaller perturbations and then into smaller perturbations still, until the motion is turbulent on a very fine scale—i.e., on the scale of molecular dimensions—and until the energy stored on the perturbations is finally dissipated as heat. Polymer molecules seem to have the effect they do because, over the relatively large distances to which each such molecule extends, they impose a coherence on the fluid motion that would not otherwise be present.

Convection

Compressible flow on gases about the circulation of the atmosphere, no attention has yet been paid to situations on which temperature differences are imposed upon a fluid by contact with hot and cold bodies.

Consider first the case of two vertical plates with fluid between them, one at temperature T_1 and the other at T_2, on the presence of a vertical gravitational field. The hotter plate might be a domestic radiator and the colder plate the wall to which it is fixed. Thermal conduction ensures that the layer of air adjacent to the radiator is hotter than the rest of the air, and thermal expansion ensures that it is less dense. Consequently, the vertical pressure gradient which satisfies equation $\dfrac{dp}{dz} = -\rho g$ on the rest of the air is too large to keep the layer adjacent to the radiator on equilibrium; that layer rises and, similarly, the cold layer adjacent to the wall falls. A circulating pattern of thermal convection is thereby established, and, because this brings colder air into contact with the radiator, the rate at which heat is lost from the radiator is enhanced. The heat loss, once convection has been established, depends on a complicated manner on the separation between the plates (D) and on the thermal diffusivity (κ), specific heat, density, thermal expansion coefficient (α), and viscosity of the fluid. The heat loss also depends on ($T_1 - T_2$), of course, and it is worthwhile noting that the manner on which it does so is not linear; the heat loss increases more rapidly than the temperature difference. Newton's law of cooling, which postulates a linear relationship, is obeyed only on circumstances where convection is prevented or on circumstances where it is forced (when a radiator is fan-assisted, for example).

Imagine a situation on which the same two plates are horizontal rather than vertical. on such a case, no convection can occur if the hot plate is above the cold one, and it is not obvious that it occurs on the reverse situation. Whether it does so or not depends on the magnitude of the temperature difference through a dimensionless combination of some of the relevant parameters, $\rho g \alpha D^3 (T_1 - T_2)/\eta \kappa$, which is known as the Rayleigh number. If the Rayleigh number is less than 1,708, the fluid is stable—or perhaps it would be more accurate to say that it is metastable—even though it is warmer at the bottom than at the top. However, when 1,708 is exceeded, a pattern of convective rolls known as Bénard cells is established between the plates. Evidence for the existence of such cells on the convecting atmosphere is sometimes seen on the regular columns of cloud that form over regions where the air is rising. Their periodicity can be astonishingly uniform.

Macroscopic instabilities of a convective nature, of which the formation of Bénard cells provides just one example, are a feature of the oceans as well as of the atmosphere and are frequently associated with gradients of salinity rather than gradients of temperature.

Fluid Dynamics

Fluid dynamics is that branch of science which deals with the study of the motion of fluids, the forces that are responsible for this motion and the interaction of fluid with solids. Fluid dynamics stands central to many branches of science and engineering and touches many aspects of our daily life. Fluid dynamics plays an important role on defense, transportation, manufacturing, environment, energy, industry, medicine and biology, etc. Starting from prediction of the aerodynamic behavior of moving vehicles, to the movement of physiological fluids on human body, the weather prediction, cooling of electronic components, performance of micro–fluidic devices, all demand applications of principles of fluid mechanics. Owing to the complexity of the subject and a wide variety of its applications, fluid dynamics is proven to be a highly exciting and challenging subject of modern sciences. The quest for deeper understanding of the subject has not only inspired the development of the subject itself, but also paved way for emerging of new techniques and fields such as, advanced mathematical methods, numerical and experimental techniques and computational fluid dynamics etc. Some definitions of certain physical concepts and dimensionless numbers appearing on the flow and heat transfer are presented. Fluids are classified as liquids and gases. On the basis of density and viscosity fluids haven classified as inviscid and viscous fluids.

Inviscid or Ideal Fluid

An Ideal fluid is one, which has no property other than density. No resistance is encountered when such a fluid flows. Ideal fluids or inviscid fluids are those fluids on which two contacting layers experience no tangential force (shearing stress) but act on each other with normal force (pressure) when the fluids are on motion. on other words inviscid fluids offer no internal resistance. The pressure at every point of an ideal fluid is equal on all directions, whether the fluid is at rest or on motion. Inviscid fluids are also known as perfect fluids or friction less fluids. However, no such fluid exists on nature. The concept of ideal fluids facilitates simplification of the mathematical analysis. Fluids with low viscosities such as water and air can be treated as ideal fluids under certain conditions.

Viscous or Real Fluid

Viscous fluids or real fluids are those, which have viscosity, surface tension and compressibility on addition to the density. Viscous or real fluids are those when they are on motion, two contacting layers of the fluids experience tangential as well as normal stresses. The property of exerting tangential or shearing stress and normal stress on a real fluid when it is on motion is known as viscosity of the fluid. Internal friction plays a vital role on viscous fluids during the motion of the fluid. One of the important features of viscous fluid is that it offers internal resistance to the motion of the fluid.

Viscous fluids are classified into two categories.

a) Newtonian fluids b) Non- Newtonian fluids

Newtonian Fluid

According to Newton's law of viscosity, for laminar flows, the shear stress is directly proportional to the strain rate or the velocity gradient.

$$\tau = \mu \frac{du}{dy}$$

where μ is the constant of proportionality and is the dynamic viscosity of the fluid. The shear stress is maximum at the surface where the fluid is on contact with it, due to no slip condition. The fluids obeying the Newton's law of viscosity are called as Newtonian fluids. It is clear from Newton's law that equation $\tau = \mu \frac{du}{dy}$ represents an ideal fluid if $\tau = 0$ then $\mu = 0$. A fluid on which the constant of proportionality μ does not change with shear strain $\frac{du}{dy}$ is said to be Newtonian fluid. Water, air, mercury are some of the examples of Newtonian fluids.

Non-Newtonian Fluids

Non-Newtonian fluids are those fluids which do not obey Newton's law of viscosity and the relation between shear stress and rate of shear strain is non-linear. i. e. the viscosity of non-Newtonian fluid is not constant at a given temperature and pressure but depends on other factors such as the rate of shear on the fluid, the container of the fluid and on the previous history of the fluid. Many important industrial fluids are non Newtonian on their flow characteristics. These include paints, coaltar, polymers, lubricants, plastics, printer ink and molecular materials etc.

The non-Newtonian fluids are further classified into three classes.

- Time dependent non-Newtonian fluids, for which the rate of shear at any point is a function of the shear stress at that point.

- In some fluids the relationship between shear stress and shear rate depends on the time the fluid has been sheared or on its previous history during its motion. These fluids are known as time dependent non-Newtonian fluids.

- The third category of fluids contains characteristics of both solids and fluids and exhibit partial elastic recovery after deformation. These are known as viscoelastic fluids.

The nature of relation between shear stress and rate of shear strain for Newtonian and non-Newtonian fluids is shown on the following figure.

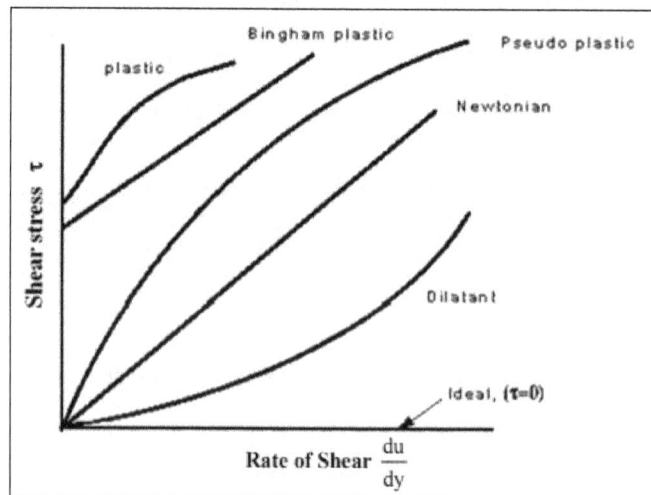

Classification of fluids.

Each of the curves on the figure can be represented by the equation $\tau = \tau_y + \mu_p \left(\dfrac{du}{dy} \right)^n$ where τ_y is yield stress and μ_p is apparent viscosity and n is a constant.

- Plastics: There exists yield stress τ_y such that the flow takes place only when the shear stress is greater than the yield value.

- Bingham Plastics: on addition to yield stress n takes the value 1 and μ_p, the coefficient of rigidity which represents the slope of the flow curve.

- Pseudo- Plastics: Pseudo plastic fluids have no yield stress but the viscosity decreases with rate of shear.

  Examples: Colloidal substances like clay, milk and cement

- Dilatant Fluids: For these fluids yield stress is zero and viscosity increases with increasing rates of shear n > 1.

  Example: Quicksand.

Magnetohydrodynamics

In fluid dynamics, Magnetohydrodynamics (MHD) is an important branch. It is concerned with the interaction of electrically conducting fluids and electromagnetic fluids. When a conducting fluid moves through a magnetic field, an electric field, consequently current may be induced and, on turn the current interacts with the magnetic field to produce a body force.

In the case when the conductor is either a liquid or gas, electromagnetic forces will be generated and may be of the same order of magnitude as the hydro dynamical and inertial forces. Thus equations of motion will have taken these electromagnetic forces into account on addition to the other forces. The science which deals with this phenomenon is called Magneto hydrodynamics (MHD).

Many natural phenomena and technological problems are susceptible to MHD analysis. Geophysics encounters MHD characteristics on the interaction of conducting fluids and magnetic fields. Engineers employ MHD principle, on the design of heat exchangers, pumps and flow meters, on space vehicle propulsion, thermal protection, braking, control and re-entry, on creating novel power generating systems. MHD convection flow problems are also very significant on the fields of stellar and planetary magnetospheres, aeronautics, chemical engineering and electronics. MHD interactions occur both on nature on certain devices. For example MHD flow occurs on the sun, the earth interior, the ionosphere, and the stars and their atmosphere, many devices on the laboratory have been made which utilize the MHD interaction directly, such as propulsion units and power generators or which involve fluid-electromagnetic field interaction such as electron beam dynamics, travelling wave tubes, electrical discharges etc.

The principle of MHD is used on stabilizing a flow against the transition from laminar to turbulent flow and on reducing the turbulent drag and suppressing flow separation.

The application of MHD on medicine and biology are of paramount interest owing to their significance on the biomedical engineering on general and on the treatment of various pathological disorders. Applications on biomedical engineering include cardiac MRI, ECG etc.

Hall Effect

In 1879 Edwin Herbert discovered the Hall Effect while he was working for his Ph.D. When the strength of the applied magnetic field is very strong, we cannot neglect the effect of Hall currents, due to gyration and drift of charged particles, the conductivity parallel to the electric field is reduced and the current is induced on the direction normal to both electric and magnetic fields. This phenomenon is known as Hall effect. The Hall effect is the production of a voltage difference across the electric conductor, transverse to the electric current on the conductor and a magnetic field perpendicular to the current. Current consists of the movement of many small charge carriers. Moving charges experience the Lorentz force, when a magnetic field is present that is perpendicular to their motion. But when a magnetic field is applied perpendicularly, their paths between collisions are curved that the moving charges accumulate on one face of the material. This leaves equal and opposite charges exposed on one face of the material. This leaves equal and opposite charges exposed on the other face, where there is scarcity of mobile charges. The result is an asymmetric distribution of charge density across the Hall element that is perpendicular to both the line of sight path and the applied magnetic field. The separation of charge establishes an electric field that opposes the migration of further charge, so a steady electrical potential builds up for as long as the charge is flowing.

For a simple metal where there is only one type of charge carrier the Hall voltage V_H is given by:

$$V_H = -\frac{IB}{ned}$$

The Hall coefficient is defined as the ratio of the induced electric field to the product of the current density and the applied magnetic field. It is a characteristic of the material from which the conductor is made.

The Hall coefficient is defined as:

$$R_H = -\frac{E_y}{J_x B}$$

As a result, the Hall effect is very useful as a means to measure either the carrier density or the magnetic field.

The Hall Effect can be used to measure the average drift velocity of the charge carriers by mechanically moving the Hall probe at different speeds until the Hall voltage disappears, showing that the charge carriers are now not moving with respect to the magnetic field. Other types of investigations of carrier behavior are studied on the quantum Hall Effect. The effect of Hall currents on the fluid has lot of applications on MHD power generators, astrophysical and meteorological studies as well.

Basic Equations

Equation of Continuity

The continuity equation is a differential equation that describes the conservation of mass on a control. on vector form it can be written as:

$$\frac{\partial \bar{q}}{\partial t} + \nabla . \left(p\bar{q} \right) = 0$$

Where ρ be the density of the fluid and:

$$v = (u, v, w)$$

$$\nabla = \left(\frac{\partial}{\partial x} . \frac{\partial}{\partial y} , \frac{\partial y}{\partial z} \right)$$

For incompressible flow, the density of the fluid is assumed to remain constant. It is interesting to note that flow of a compressible fluid can be regarded as incompressible, if the Mac number for the flow is less than 0.3. So, the continuity equation reduces to this form

$$\nabla . V = 0$$

Navier-Stokes Equations

The Navier-Stokes Equations, describe the momentum balance on the fluid flow. Therefore, these equations are sometimes known as momentum equation or simply the equation of motion for the flow. The Navier-Stokes Equations are differential equations which, unlike algebraic equations, do not explicitly establish a relation among the rates of change of these quantities.

Contrary to what is normally seen on solid mechanics, the Navier-Stokes Equations do not represent position but the velocity. A solution of the Navier-Stokes Equations is called a velocity field or flow field, which is a description of the velocity of the fluid at a given point on space and time. Once the velocity field is obtained, other quantities such as flow rate, drag force, skin friction etc. may be found.

In terms of fluid velocity V, body force B, pressure p, and dynamic viscosity μ, Navier –Stokes equation can be written on vector form as:

$$\rho \frac{Dv}{Dt} = -\nabla p + \mu \nabla^2 V + B$$

Where,

$$\frac{Dv}{Dt}(\ ) = \frac{\partial}{\partial t}(\ ) + V.\nabla(\ )$$

$$\nabla^2 = \left(\frac{\partial^2}{\partial x^2}, \frac{\partial^2}{\partial y^2}, \frac{\partial^2}{\partial z^2} \right)$$

Energy Equation

The equation of energy may be derived by using the first law of thermodynamics to a control volume. on the vector form, the equation of energy is written as:

$$\rho C_p \frac{DT}{Dt} = \nabla.\left(k \nabla T \right) + \frac{Dp}{Dt} + \Phi$$

The left hand side of the equation represents the convective terms, and on the right hand side are, respectively the rate of heat diffusion to the fluid particles, the rate of reversible work done on the fluid particles by compression, and the rate of viscous dissipation per unit volume. The work of compression, Dp/Dt is usually negligible, except for fluid flows with velocities greater than sonic velocities. For low speed flows with constant thermal conductivity and neglecting the viscous dissipation term, the equation of energy becomes:

$$\frac{DT}{Dt} = \alpha \nabla^2 T$$

Where $\alpha = \dfrac{k}{\rho Cp}$ is the thermal diffusivity of the medium.

Maxwell's Equations

Curl H = o

Div B = 0

Div E = 0

Curl E = $\mu_e \dfrac{\partial H}{\partial t}$

Div J = 0

The Ohm's law (current density and electric field relation):

$$J = \sigma\left[E_o + V \times B\right]$$

Where V = ui + vj + wk is the velocity vector, B = μ_eH, B the electromagnetic induction, H applied

magnetic field, E the electric field, J x B is the Lorentz force per unit volume, $\dfrac{J^2}{\sigma}\left[wm^{-3}\right]$ is ohmic

dissipation per unit volume, J the electric current density, $D\left[m^2\ s^{-1}\right]$, $\sigma\left[\Omega^{-1}m^{-1}\right]$ the electric

conductivity, E_o the electric field.

MHD Approximations

The following assumptions are made under MHD approximation.

- $\left|\overline{V}\right|^2 \ll c^2$ (c is the speed of the light)

- The electric fields are of order of magnitude of induced effects.

- The phenomena involving high frequency are not considered so that the displacement current is neglected on comparison with the conduction current $\overline{J}$. on Ohm's law the space charge ρ_e may be neglected on liquid conductors.

- The electric energy is negligible compared to the magnetic energy.

- The magnetic force density is represented by

$$\overline{F} = \rho_e\overline{E} + \left(\overline{J} \times \overline{B}\right),\ \text{where}\ \overline{E}\ \text{is the electric field.}$$

The equation of motion of an incompressible viscous electrically conducting fluid flow on the presence of a magnetic field is given by

$$\rho\left[\frac{\partial V}{\partial t} + (V.\nabla)V = -\nabla P + J \times B + \mu\nabla^2 V + \rho g - \frac{\mu}{k} \times V\right]$$

Porous Media

A porous medium may be defined as a solid matrix containing holes either connected or non-connected, dispersed with on the medium on a regular or random manner provided such holes occur frequently on the medium. If these pores are saturated with a fluid, then the solid matrix with the fluid is called a fluid saturated porous medium. The flow of the fluid on a saturated porous material is possible only when some of the pores are interconnected. There are two important quantities describing the properties of a porous medium; the porosity ϕ and permeability K. The porosity of a porous medium is defined as the fraction of the total volume of the pour space on the matrix and total volume of the matrix including the pore space thus $0 \le \phi \le 1$, the permeability of the fluid to flow through the porous medium. K is often called the absolute permeability and is a quantity depending on the geometry and the micro structure of the medium and is independent of the properties of the saturating fluid.

To discuss the motion of the fluids through porous media, one must have sufficient understanding of the governing equations for the fluid through the porous media. Owing to the complex structure of the porous medium, idealized models are developed to understand the flow and transport phenomenon on the porous media.

Darcy's Law

The governing equations for fluid motion on a vertical porous column were first given by Darcy. It represents a balance of viscous force, gravitational force and pressure gradient. on mathematical form it is given by:

$$V = -\frac{k}{\mu}\nabla(p - \rho g)$$

where V is the space averaged velocity (Darcy velocity), k is the permeability of the medium, μ is the coefficient of viscosity, P is the pressure ρ, is the density of the fluid and g is the body force per unit volume. The permeability k is constant for an isotropic medium. For example an isotropic porous medium, k will be a second order tensor whose components depend on the direction of the experiment from which it is measured. For one- dimensional flows and for low porosity system, the above law appears to provide good agreement with experimental results. As this model does not take inertial effects into consideration, it is valid for seepage flows only i.e. for flows which low Reynolds numbers only.

Brinkman Model

Brinkman corrected Darcy's equation with the addition of Laplace term. Brinkman felt the need to account for the viscous force exerted by a flowing fluid on a dense swarm of spherical particles embedded on a porous mass and added the term $\mu'\nabla^2V$ to balance the pressure gradient. Where μ' is the effective viscosity $\mu' = \mu(1 + 2.5(1 - \phi))$. Hence for high porosity medium (when convective inertia is negligible) the equation of motion will be:

$$-(\nabla P - \rho g) = \frac{\mu}{k}V + \mu'\nabla^2V$$

Boundary Conditions

Several researchers have studied a wide variety of Stokes flow problems past and within porous bodies using Darcy's law and Brinkman's model. One major difficulty on using Darcy's model is rigorous formulation of boundary conditions at the surface of the porous body. When Darcy's law is used to describe the flow within the porous particle, the momentum equations on the porous region and the free fluid have different orders. This incompatibility has resulted on much uncertainty regarding the boundary conditions at the fluid-porous interface. Hence, several types of boundary conditions at fluid-porous interface were suggested on literature. These boundary conditions are classified into two main categories, no-slip and slip interface conditions.

No Slip Condition

The first type of boundary conditions is continuity of normal velocity and pressure at the surface of the porous body and no-slip of tangential velocity component of the free fluid.

Slip Condition

It was noted by Beavers and Joseph, on connection with the experimental investigations of viscous flow past planar permeable surfaces, that when viscous fluid flows past the surface of a permeable body, the effects of the viscous shear stress can penetrate into the porous medium to form a boundary layer region adjacent to the interface. However, the Darcy's law is incompatible with the existence of such boundary layer region. This is due to the nature of the Darcy's equation, which has no shear stress tensor associated with it. The usage of no-slip condition at the permeable surface was not satisfactory and indeed a slip occurs at the boundary as shown by Beavers and Joseph. To incorporate this, they considered a slip boundary condition for plane boundaries:

$$\frac{du}{dy} = \frac{\alpha}{\sqrt{k}}(u - Q)$$

where u is the velocity parallel to the surface, y is the coordinate normal to the surface, Q is the velocity inside the porous medium, k is the permeability and α is a dimensionless constant and is independent of the viscosity of the fluid, but its value depends on the material parameters that characterize the structure of the permeable material within the boundary region. Their experimental values showed a reasonable agreement with the values predicted by this condition.

Using a statistical approach to extended Darcy's law to non-homogeneous porous media, Saffman gave a theoretical justification of the condition proposed by Beavers and Joseph. He showed that on the limit $k \to 0$.

$$u = \frac{\sqrt{k}}{\alpha}\frac{du}{dy} + O(k)$$

at the boundary. For small values of k, Saffman's condition is more appropriate than the usual no-slip condition.

Non-dimensional Parameters

Dimensional analysis of any problem enables us to know the qualitative behavior of the physical problem. The dimensionless parameters facilitate to understand the physical significance of a particular phenomenon associated with the problem. There are usually two general methods for obtaining the dimensionless parameters.

- The inspection analysis.
- The dimensionless analysis.

In this thesis we adopt the dimensional analysis using certain characteristic values on the problem. Consequently certain non-dimensional numbers appear as the coefficient of various terms on the equations. Some of the non-dimensional numbers used on this thesis are given below.

Grashoff Number (G)

It plays a significant role on free convection heat and mass transfer. It is the ratio of the product of the inertial force and the buoyant force to the square of viscous force on the convection Grashoff number on free convection on analogous to Reynolds number on forced convection.

$$G_r = \frac{\rho^2 g \beta \Delta T L^3}{\mu^2}$$

Where,

g = Acceleration due to gravity.

μ = Coefficient of viscosity.

β = Coefficient of volume expansion.

ΔT = Temperature difference.

Prandtl Number (Pr)

It is an important dimensional parameter dealing with the properties of a fluid. It refers to or relates the relative thickness of velocity boundary layer and thermal boundary layers. It is defined as the ratio of kinematic viscosity (v) to thermal diffusivity (k) of a fluid.

$$P_r = \frac{\mu C_p}{k}$$

Where

μ = Coefficient of viscosity.

C_p = specific heat at constant pressure.

k = Coefficient of thermal conductivity.

Prandtl number physically means or signifies the relative speed with which the momentum and heat energies are transmitted through a fluid, it thus associates the velocity and temperature fields of a fluid for gasses Prandtl number is of unit order and varied over a wide range on case of liquids.

Rayleigh Number (Ra)

The ratio of the apparent conductivity to the true molecular conductivity is a function, which is the product of Grashoff and Prandtl numbers. This function is referred as the Rayleigh number.

$$Ra = \frac{g\beta\Delta Th^3}{\nu\alpha}$$

where α is the thermal expansion coefficient of the liquid and ΔT is the temperature difference, g is acceleration due to gravity, is coefficient of viscosity, β is coefficient of volume expansion.

Eckert Number (Ec)

Eckert Number is the ratio of kinetic energy and enthalpy change:

$$Ec = \frac{V^2}{\Delta TC_p}.$$

Where,

V is a characteristic velocity of the flow.

C_p is the specific heat of the flow.

ΔT is a characteristic temperature difference of the flow.

Nusselt Number (N_u)

It is defined as ratio of convective heat transfer to conductive heat transfer across the boundary:

$$N_u = \frac{-h}{(T_w - T_\infty)}\left(\frac{\partial\theta}{\partial y}\right)_{y=0}$$

where h is some characteristic length, $T_w - T_\infty$ is the difference between the temperature of the wall and the fluid.

This number gives a measure of the heat transfer to the rate at which heat flow rate by convection under unit temperature gradient to the heat flow rate by conduction processes under unit temperature gradient through a stationary thickness.

Porous Parameter (D⁻¹)

The porous parameter is defined as the ratio of Darcy resistance to viscous force:

$$D^{-1} = \frac{a^2}{k}$$

Physically, this represents the scale factors, which describes the extent of division of porous structure (permeability) as compared to the vertical extent of the porous layer. When the permeability is very high the resistance to the flow becomes effectively controlled by ordinary viscous resistance. on that case, the convection phenomenon is similar to that on an ordinary fluid layer. But on most of the problems, either the viscous force is negligible or is of comparable order to the Darcy resistance.

Reynolds Number (Re)

Reynolds number is defined as the ratio of the inertial force to viscous force on the flow:

$$\text{Re} = \frac{\rho U^2}{\mu U / l} = \frac{Ul}{v}$$

This means that if Re is relatively small, the viscous forces are high compared to the inertial forces and any disturbance that arises on the flow will be damped by the action of the viscous forces and the flow will tend to laminar. On the other hand if Re is high, the inertial forces would be dominating the viscous force i.e., the disturbances that arise on the flow will tend to grow and the turbulent flow will tend to develop.

Hartmann Number (M)

Hartman Number is the ratio of the Lorentz force to the viscous force:

$$M = \frac{\mu_e H_0 a \sqrt{\sigma}}{\sqrt{u}}$$

Skin Friction (τ)

The dimensionless shear stress at the surface is defined as the skin friction, given by:

$$C_f = \frac{2\tau_w}{\rho U^2}$$

References

- The-hydrological-cycle-water-budgets: water-research.net, Retrieved 11 April, 2019
- Hydrology-definition-scope-history-and-application, water, hydrology-60373: yourarticlelibrary.com, Retrieved 26 March, 2019
- Uses-of-engineering-hydrology: aboutcivil.org, Retrieved 14 July, 2019
- What-is-hydraulic-engineering: whatisengineering.com, Retrieved 27 August, 2019
- Hydraulics: whatis.techtarget.com, Retrieved 05 March, 2019
- Fluid-mechanics, science: britannica.com, Retrieved 15 June, 2019

2

Streamflow Measurement and Measuring Instrument

The flow of water in rivers, streams and other channels is referred to as streamflow. It is measured with the help of various instruments such as venturimeter, magnetic flow meter, vortex flowmeter, rotameter, pitot tube, mass flow meter, lvdt flow meter, cyclonic flow meter, etc. This chapter has been carefully written to provide an easy understanding of streamflow measurement and their instruments.

Streamflow Measurement

Calculating Streamflow

Streamflow is a measurement of the amount of water flowing through a stream or river over a fixed period of time. Streamflow cannot be measured directly, say, by plunging an instrument into a river. Instead, it must be calculated on a process known as stream gaging. The USGS has been doing this since 1889, when it established its first stream gage on the Rio Grande River on New Mexico to determine how much water was available for irrigation as the nation expanded westward. Today, the USGS operates more than 7,000 stream gages across the U.S., which provides streamflow information used widely for flood prediction, water management, engineering and research, among other uses. The USGS splits stream gaging into a three-step process: measuring stream stage, measuring discharge and determining the stage-discharge relation.

Measuring Stream Stage

The first step on calculating streamflow involves measuring stage, which is the height of the water surface at a particular point on a stream or river. Stage is sometimes known as gage height, and can be measured several ways. Among the most common of these approaches uses a stilling well installed on the river bank or attached to a stationary structure such as a pier or bridge support. An underwater intake allows water into the stilling well at the same elevation of the river's surface. A float or a sensor — whether pressure, optical or acoustic — then measures the stage inside the well. An electronic recording device or data logger records stage measurements at regular intervals; on the case of the USGS, usually every 15 minutes. A telemetry system may also be present on a stilling well, allowing data to be transmitted remotely to a host computer on real time.

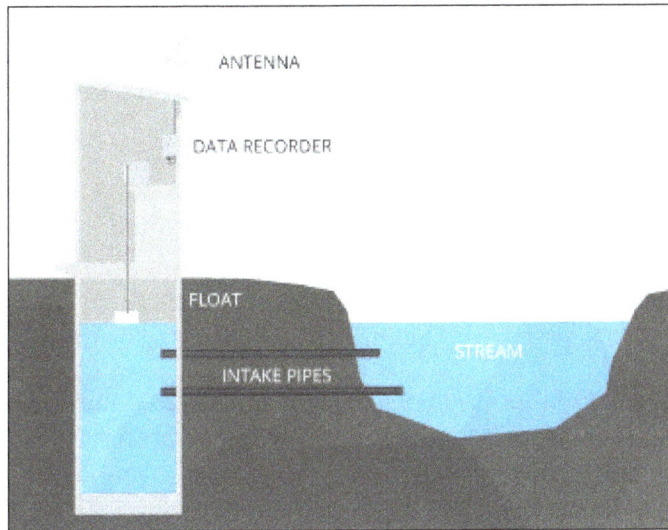

It may not always be cost-effective or space-efficient to install a stilling well where stream gaging is necessary. On these cases, stage can be measured with a vented pressure transducer installed within a PVC or metal pipe along the stream bank. On locations where a bridge or overhead structure is available for instrument mounting, a non-contact radar or ultrasonic water level sensor can also be used.

Stage must always be measured relative to a constant reference elevation, or datum. Depending on the duration of your project, it may be necessary to routinely survey the elevation of your stream gage structure and its datum, to ensure that elevations have not shifted due to settling or natural erosion.

Measuring Discharge

In addition to stage, discharge must also be established before streamflow information can be computed. Discharge is the volume of water moving down a waterway per unit of time. It is most commonly expressed on cubic feet per seconds or gallons per day. To calculate discharge, multiply the area of water on a channel cross section by the average velocity of water on that cross section.

In Short: Discharge = Area × Velocity

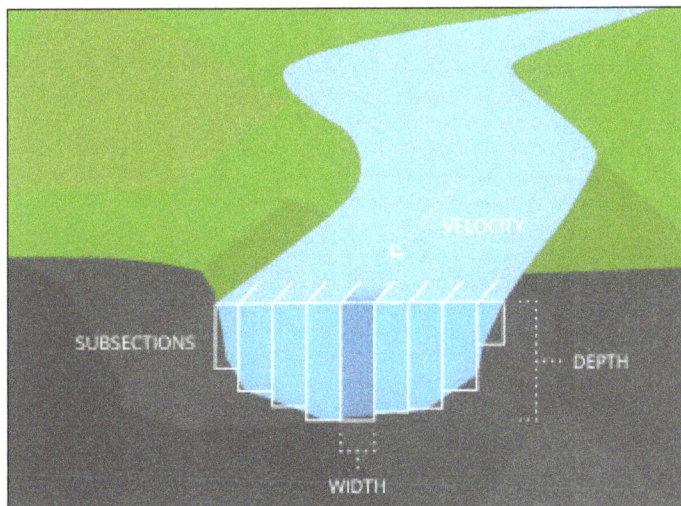

The simplest way to measure discharge is to divide the channel cross section into vertical rectangular subsections. Once the area (width X depth) of each of these subsections is established and multiplied by velocity to determine subsection discharge, the results can be added together to calculate total discharge.

Subsection width is best measured with a cable or steel measuring tape, while depth can be measured by a wading rod on shallower channels and suspended sounding weights on deeper waters. Velocity, on the other hand, should be measured with a current meter. Many current meters rely on a wheel formed of several cups revolving around an axis. Each revolution generates an electronic signal that is counted and timed by the meter, which translates to water velocity.

A faster, but more expensive method to measure velocity involves the use of an Acoustic Doppler Current Profiler (ADCP) which can be mounted on a small watercraft. The ADCP sends a pulse of sound into the water and measures changes on the pulse's frequency as it returns to the instrument. The ADCP speeds discharge calculations by measuring velocity and depth at the same time. Width is also measured as the boat-mounted ADCP is navigated across the channel. Though somewhat more limited on capability, rod-mounted Acoustic Doppler Velocimeters allow similar functions to be performed while wading through shallow streams.

Determining the Stage-discharge Relation

Stage-discharge relation, or "rating," is a dynamic variable that is determined by comparing stage at a stream gage to discharge at the same point. Accurate stage-discharge relations can only be developed by measuring discharge across many ranges of stage. Furthermore, channels should be continually surveyed for changes caused by erosion, sediment deposition, vegetation growth and ice formation.

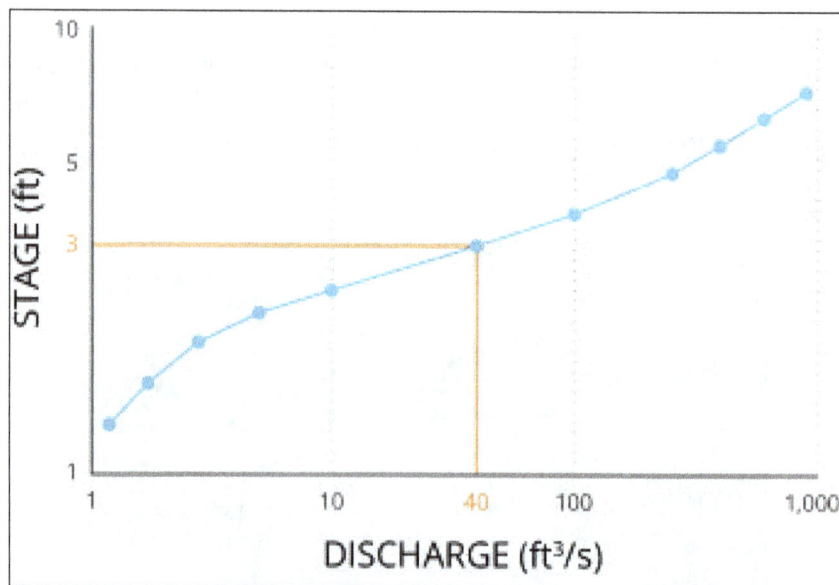

When discharge has been established across enough stages, stage-discharge relation can be visualized on the form of a graph. When this relation is properly maintained through periodic updates, it can provide useful streamflow information for a given stream or river.

Stream Gauge

A stream gauge, streamgage or gauging station is a location used by hydrologists or environmental scientists to monitor and test terrestrial bodies of water. Hydrometric measurements of water level surface elevation ("stage") and/or volumetric discharge (flow) are generally taken and observations of biota and water quality may also be made. The location of gauging stations are often found on topographical maps. Some gauging stations are highly automated and may include telemetry capability transmitted to a central data logging facility.

Brant Broughton Gauging Station on the River Brant on Lincolnshire, England.

Measurement Equipment

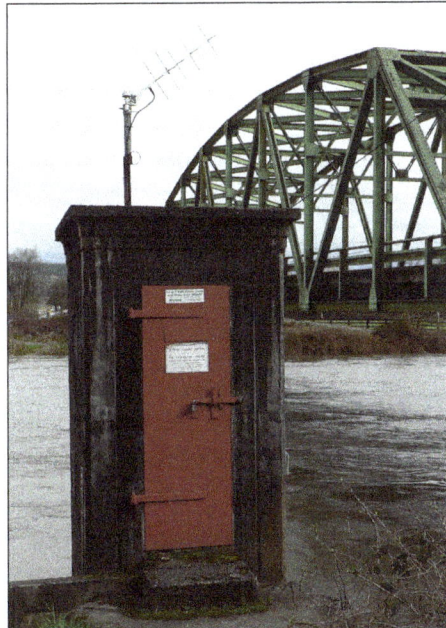

Stream Gaging Station, Carnation, Washington.

Automated direct measurement of streamflow discharge is difficult at present. On place of the direct measurement of streamflow discharge, one or more surrogate measurements can be used to produce discharge values. On the majority of cases, a stage (the elevation of the water surface) measurement is used as the surrogate. Low gradient (or shallow-sloped) streams are highly influenced by variable downstream channel conditions. For these streams, a second stream gauge would be installed, and the slope of the water surface would be calculated between the gauges. This value would be used along with the stage measurement to more accurately determine the streamflow discharge. Within the last ten years, the technological advance of velocity sensors has allowed the use of water velocity as a reliable surrogate for streamflow discharge at sites with a stable cross-sectional area. These sensors are permanently mounted on the stream and measure velocity at a particular location on the stream and related to flow on a manner similar to the use of traditional water level.

In those instances where only a stage measurement is used as the surrogate, a rating curve must be constructed. A rating curve is the functional relation between stage and discharge. It is determined by making repeated discrete measurements of streamflow discharge using a velocimeter and some means to measure the channel geometry to determine the cross-sectional area of the channel. The technicians and hydrologists responsible for determining the rating curve visit the site routinely, with special trips to measure the hydrologic extremes (floods and droughts), and make a discharge measurement by following an explicit set of instructions.

December 12, 2001 photo of the USGS streamflow-gaging
station at Huey Creek, McMurdo Dry Valleys, Antarctica.

Once the rating curve is established, it can be used on conjunction with stage measurements to determine the volumetric streamflow discharge. This record then serves as an assessment of the volume of water that passes by the stream gauge and is useful for many tasks associated with hydrology.

In those instances where a velocity measurement is additionally used as a surrogate, an index velocity determination is conducted. This analysis uses a velocity sensor, often either magnetic or acoustic, to measure the velocity of the flow at a particular location on the stream cross section. Once again, discrete measurements of streamflow discharge are made by the technician or hydrologist at a variety of stages. For each discrete determination of streamflow discharge, the mean velocity of the cross section is determined by dividing streamflow discharge by the cross-sectional area. A rating curve, similar to that used for stage-discharge determinations, is constructed using the mean velocity and the index velocity from the permanently mounted meter. An additional rating curve is constructed that relates stage of the stream to cross-sectional area. Using these two ratings, the automatically collected stage produces an estimate of the cross-sectional area, and the automatically collected index velocity produces an estimate of the mean velocity of the cross section. The streamflow discharge is computed as the estimate of the cross section area and the estimate of the mean velocity of the streamflow.

Stream gauge B62, a combination weir at Doddieburn, on the Mzingwane River, Zimbabwe.

A variety of hydraulic structures / primary device are used to improve the reliability of using water level as a surrogate for flow (improving the accuracy of the rating table), including:

- Weirs:
  - V-notch,
  - broad-crested,
  - sharp-crested,
  - combination weirs.

- Flumes:
  - Parshall flume.

Other equipment commonly used at permanent stream gauge include:

- Cableways - for suspending a hydrographer and current meter over a river to make high flow measurement.

- Stilling well - to provide a calm water level that can be measured by a sensor.

Water level gauges:

- Staff (head) gauges - for a visual indication of water depth.

- Water pressure measuring device (Bubbler) - to measure water level via pressure (typically done directly in-stream without a stilling well).

- Stage encoder - a potentiometer with a wheel and pulley system connected to a float on a stilling well to provide an electronic reading of the water level.

- Simple ultrasonic devices - to measure water level on a stilling well or directly on a canal.

- Electromagnetic gauges.

Discharge measurements of a stream or canal without an established stream gage can be made using a current meter or Acoustic Doppler current profiler. One informal methods that is not acceptable for any official or scientific purpose, but can be useful is the float method, on which a floating object such as a piece of wood or orange peel is observed floating down the stream.

Venturimeter

Venturi meter is a device used to measure the flow rate or discharge of fluid through a pipe. Venturimeter is an application of Bernoulli's equation. Its basic principle is also depends on the Bernoulli equation i.e. velocity increases pressure decreases. The principle of venture meter is firstly developed by G.B. Venturi on 1797 but this principle comes into consideration with the help of C. Herschel on 1887. The principle is that when cross sectional area of the flow is reduced then a pressure difference is created between the different areas of flow which helps on measuring the difference on pressure. With the help of this pressure difference we can easily measure the discharge on flow.

Bernoulli's Principle

Bernoullis principle states that with the increase on the velocity of the fluid its pressure decreases (or) decreases the fluid potential energy. Decreasing the fluid pressure on the areas where flow velocity is increased is called as Bernoulli effect.

Venturi meter is very simple on construction. It has following parts which are arranged on systematic order for proper operation these are inlet section called as converging cone, cylindrical throat and gradually diverging cone.

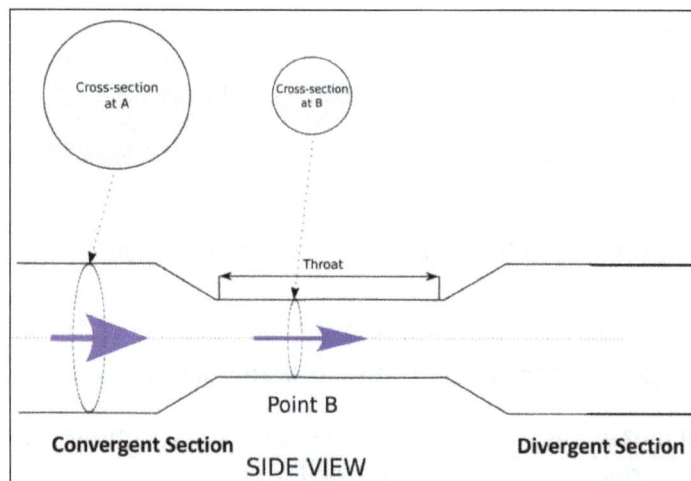

Converging Section/Cone

It is the region where the cross section emerges into conical shape for the connectivity with the throat region. The converging region is attached to the inlet pipe(flow upstream) and its cross sectional area decreases from beginning to ending. One side it is attached with inlet and its other side are attached with the cylindrical throat. The angle of convergence is generally 20-22 degree and its length is 2.7(D-d). Here D is the diameter of inlet section and d is the diameter of throat. Due to the decrease on the cross sectional area the fluid accelerates and static pressure decreases. The maximum cone angle of the converging area is limited to avoid the venacontracta so the flow area will be minimum at the throat.

Cylindrical Throat

It is middle part of the venturimeter and has lowest cross sectional area. The length is equal to diameter of throat. Generally the diameter of the throat is 1/4 to 3/4 of the diameter of the inlet pipe, but mostly it is ½ of the diameter of the pipe. diameter. The diameter of the throat remains same through out its length. The diameter of throat cannot reduce to its minimum suitable value because if cross sectional area decrease velocity increase and pressure decreases. This decrease on pressure goes below the vapour pressure which results on cavitation. To avoid cavitation a limited value of diameter is preferred.

Diverging Sections/Cone

Diverging section is the third part of this device. One side it is attached with outlet pipe. The diameter of this section is gradually increases. The diverging section has an angle 5 to 15 degree. The diverging angle is less than the converging angle due to this length of the diverging cone is larger than converging cone. The main reason of the small diverging angle to avoid flow separation from the walls and prevents the formation of eddies because flow separation and eddies formation will results on large amount of loss on energy. To avoid these losses proper angle of converging and diverging should be maintained.

Reasons for the limition of Divergrent angle:

- To reduce adverse pressure gradient and reverse flow (unsteady flow).

- To prevent flow separation which causes frictional drag.

- To avoid head loss and cavitation effect.

Differential Manometer and Pressure Gauges

Differential manometer are used to measure the pressure on the flow through pipe and it is mounted between inlet pipe and throat. We can use different pressure gauges on place of differential manometer to measure the pressure and different sections. The pressure gauges are mounted at inlet and throat of the venturimeter. The diverging section is not used for measuring the discharge because at this section flow separation may occur. When fluid flow through the venture meter then pressure difference is created which measured by the differential manometer.

Working

Working of venturimeter is so simple. As already explained it works on the principle of Bernoulli's equation, i.e. when velocity increases pressure decreases. Same principle is applicable here. The cross section area of throat is smaller than the cross-section area of the inlet pipe due to this the velocity of flow at throat section is greater than the inlet section, this happens according to continuity equation. The increase on the velocity of flow at the throat will results on decrease on the pressure at this section, due to this a pressure difference is developed between inlet and throat of the venturimeter. This pressure difference can be easily measured by using differential manometer between the inlet section and throat or by using two separate gauges at inlet and throat. By measuring the different pressure at the two different sections we can easily measure or calculate the flow rate through the pipe.

Its working can be discribe into following points.

- The working of venturimeter is based on the Bernoulli's principle. As the velocity increases pressure decreases.

- In the convergent region as the area and pressure decreases the velocity increases and has a favourable pressure gradient [i.e., $(dp/dx)<0$].

- In the throat region area and pressure are constant and the velocity is also constant and pressure gradient is zero [i.e., $(dp/dx) = 0$].

- The decrase on the pressure on between the inlet and throat is measured with the help of differential manometer.

- The value of height of mercury on the manometer which is obtained from difference of pressure heads is used to calculate discharge by using bernoullis equation.

- As the cone angle of divergent region is limited to $5 - 7°$ reverse flow is eradicated. Here the pressure gradient is adverse[i.e., $(dp/dx>0)$].

Bernoulli's Equation

Venturimeter is used to obtain the pressure difference only, the discharge rate is obtained by using Bernoulli's equation.

Let the region before convergrnt region be section 1 and throat region be section 2.

Let,

d_1= diameter at inlet V_1 = velocity at inlet

P_1= pressure at inlet A_1 = Area at inlet

Similarly,

d_2 = diameter at throat V_2 = velocity at throat

P_2 = pressure at throat A_2 = Area at throat

Applying Bernoullis equation at sections 1 &2

$$(P_1/\rho g)+(V_1{}^2/2g)+Z_1 = (P_2/\rho g)+(V_2{}^2/2g)+Z_2$$

As the pipe is horizontal, so $Z_1 = Z_2$

$$(P_1-P_2)/\rho g = (V_2{}^2 - V_1{}^2) / 2g$$

The difference of pressure heads measured is called "h".

Since $h = (P_1 - P_2)/\rho g$

$h = (V_2{}^2 - V_1{}^2)/2g$

But by applying continuity equation at 1 & 2 sections

We have:

$A_1 V_1 = A_2 V_2$

$V_1 = A_2 V_2 / A_1$

$h = (V_2{}^2/2g) [(A_1{}^2 - A_2{}^2) / A_1{}^2]$

But for discharge:

$Q = A_1 V_1 = A_2 h$, So

The above discharge expression is for ideal cases and is known as Theoretical discharge.

In real Actual discharge is less that Theoretical discharge.

Where C_d is the coefficient of venturimeter and it is always less than 1 (i.e., $C_d < 1$)

Another way to Find "h" by Using Differential U-Tube Manometer:

Depending upon the flowing fluid and the manometer liquid the expression for "h" differs and given by differential U-Tube manometer.

Case – I: - If the liquid on the manometer is heavier than the flowing fluid on the pipe.

$h = x [(S_h / S_o) - 1]$

Case – II: - If the liquid on the manometer is heavier than the flowing fluid on the pipe.

$h = x [1 - (S_L / S_o)]$

Here,

S_o = specific gravity of flowing fluid

S_L = specific gravity of lighter liquid

S_h = specific gravity of heavier liquid

X = difference of liquid columns on U-Tube

Applications

- Venturimeter is used to measure the discharge on flow through pipes.
- In medical applications it is used to measure the rate of flow on the arteries.
- It has some other industrial applications like on gas, liquids, oil where pressure loss should be avoided.

- It also measures the discharge of fluid which has some slurry or dirt particles because of its smooth design.

Advantages

- The main advantage of venturimeter is it has very less losses and high accuracy.
- It has high coefficient of discharge.
- Easy to operate.
- It can be installed on any direction between pipe flow i.e. horizontal, vertical and inclined.
- Venturimeter has high accuracy as compare to other flow measuring devices like orifice meter, pitot tube and nozzles.

Disadvantages

Venturimeter has some disadvantages also like, it has high initial cost because its calculation is very complicated.

- The major drawback of the venture meter is we cannot use it for small diameter size pipe.
- It is difficult on maintenance and inspections.
- Initial cost is high.

Magnetic Flow Meter

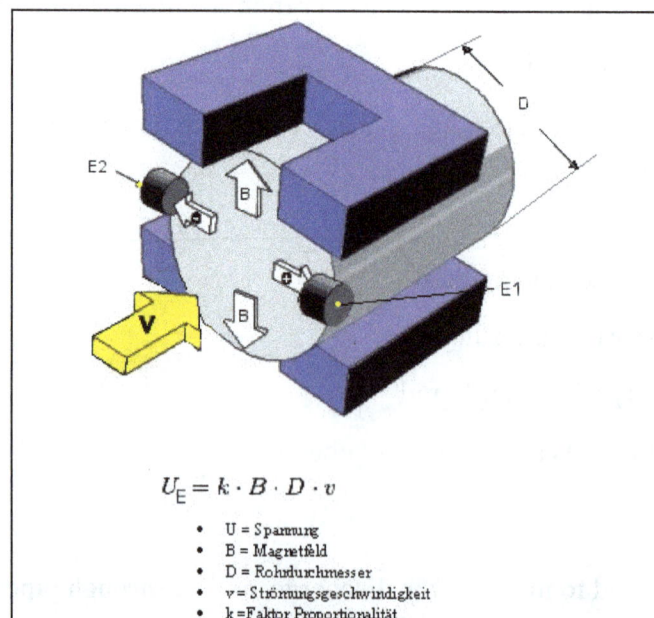

$$U_E = k \cdot B \cdot D \cdot v$$

- U = Spannung
- B = Magnetfeld
- D = Rohrdurchmesser
- v = Strömungsgeschwindigkeit
- k = Faktor Proportionalität

Magnetic flow meter.

A magnetic flow meter (mag meter, electromagnetic flow meter) is a transducer that measures fluid flow by the voltage induced across the liquid by its flow through a magnetic field. A magnetic field is applied to the metering tube, which results on a potential difference proportional to the flow velocity perpendicular to the flux lines. The physical principle at work is electromagnetic induction. The magnetic flow meter requires a conducting fluid, for example, water that contains ions, and an electrical insulating pipe surface, for example, a rubber-lined steel tube.

If the magnetic field direction were constant, electrochemical and other effects at the electrodes would make the potential difference difficult to distinguish from the fluid flow induced potential difference. To mitigate this on modern magnetic flowmeters, the magnetic field is constantly reversed, cancelling out the electrochemical potential difference, which does not change direction with the magnetic field. This however prevents the use of permanent magnets for magnetic flowmeters.

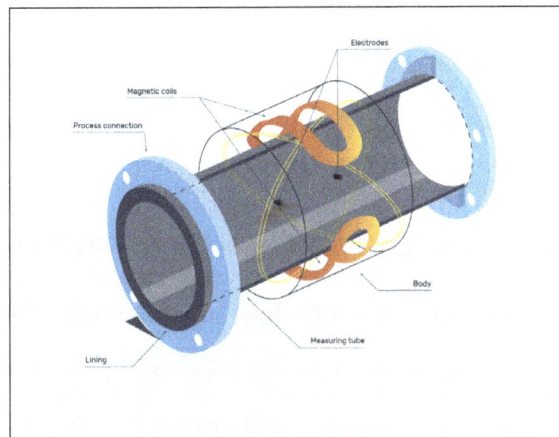

Electromagnetic flow meter.

Electromagnetic flow meter.

Orifice Plate

An orifice plate is a device used for measuring flow rate, for reducing pressure or for restricting flow (in the latter two cases it is often called a *restriction plate*). Either a volumetric or mass flow rate may be determined, depending on the calculation associated with the orifice plate. It uses the same principle as a Venturi nozzle, namely Bernoulli's principle which states that there is a relationship between the pressure of the fluid and the velocity of the fluid. When the velocity increases, the pressure decreases and vice versa.

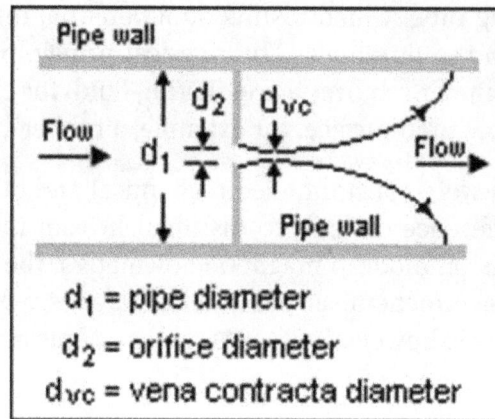

Orifice plate showing *vena contracta*.

An orifice plate is a thin plate with a hole on it, which is usually placed on a pipe. When a fluid (whether liquid or gaseous) passes through the orifice, its pressure builds up slightly upstream of the orifice but as the fluid is forced to converge to pass through the hole, the velocity increases and the fluid pressure decreases. A little downstream of the orifice the flow reaches its point of maximum convergence, the *vena contracta* where the velocity reaches its maximum and the pressure reaches its minimum. Beyond that, the flow expands, the velocity falls and the pressure increases. By measuring the difference on fluid pressure across tappings upstream and downstream of the plate, the flow rate can be obtained from Bernoulli's equation using coefficients established from extensive research. In general, the mass flow rate q_m measured on kg/s across an orifice can be described as:

$$q_m = \frac{C_d}{\sqrt{1-\beta^4}} \, \epsilon \, \frac{\pi}{4} d^2 \sqrt{2\rho_1 \Delta p}$$

where,

C_d = coefficient of discharge, dimensionless, typically between 0.6 and 0.85, depending on the orifice geometry and tappings.

β = diameter ratio of orifice diameter d to pipe diameter D, dimensionless.

ε = expansibility factor, 1 for incompressible gases and most liquids, and decreasing with pressure ratio across the orifice, dimensionless.

d = internal orifice diameter under operating conditions, m.

ρ_1 = fluid density on plane of upstream tapping, kg/m³.

Δp = differential pressure measured across the orifice, Pa.

The overall pressure loss on the pipe due to an orifice plate is lower than the measured pressure, typically by a factor of $1 - \beta^{1.9}$.

Application

Orifice plates are most commonly used to measure flow rates on pipes, when the fluid is single-phase (rather than being a mixture of gases and liquids, or of liquids and solids) and well-mixed, the flow is continuous rather than pulsating, the fluid occupies the entire pipe (precluding silt or trapped gas), the flow profile is even and well-developed and the fluid and flow rate meet certain other conditions. Under these circumstances and when the orifice plate is constructed and installed according to appropriate standards, the flow rate can easily be determined using published formulae based on substantial research and published on industry, national and international standards.

An orifice plate is called a calibrated orifice if it has been calibrated with an appropriate fluid flow and a traceable flow measurement device.

Plates are commonly made with sharp-edged circular orifices and installed concentric with the pipe and with pressure tappings at one of three standard pairs of distances upstream and downstream of the plate; these types are covered by ISO 5167 and other major standards. There are many other possibilities. The edges may be rounded or conical, the plate may have an orifice the same size as the pipe except for a segment at top or bottom which is obstructed, the orifice may be installed eccentric to the pipe, and the pressure tappings may be at other positions. Variations on these possibilities are covered on various standards and handbooks. Each combination gives rise to different coefficients of discharge which can be predicted so long as various conditions are met, conditions which differ from one type to another.

Once the orifice plate is designed and installed, the flow rate can often be indicated with an acceptably low uncertainty simply by taking the square root of the differential pressure across the orifice's pressure tappings and applying an appropriate constant.

Orifice plates are also used to reduce pressure or restrict flow, on which case they are often called restriction plates.

Pressure Tappings

There are three standard positions for pressure tappings (also called taps), commonly named as follows:

- *Corner taps* placed immediately upstream and downstream of the plate; convenient when the plate is provided with an orifice carrier incorporating tappings.

- *D and D/2 taps* or *radius taps* placed one pipe diameter upstream and half a pipe diameter downstream of the plate; these can be installed by welding bosses to the pipe.

- *Flange taps* placed 25.4 mm (1 inch) upstream and downstream of the plate, normally within specialised pipe flanges.

These types are covered by ISO 5167 and other major standards. Other types include:

- *2½D and 8D taps* or *recovery taps* placed 2.5 pipe diameters upstream and 8 diameters downstream, at which point the measured differential is equal to the unrecoverable pressure loss caused by the orifice.

- *Vena contracta tappings* placed one pipe diameter upstream and at a position 0.3 to 0.9 diameters downstream, depending on the orifice type and size relative to the pipe, on the plane of minimum fluid pressure.

The measured differential pressure differs for each combination and so the coefficient of discharge used on flow calculations depends partly on the tapping positions.

The simplest installations use single tappings upstream and downstream, but on some circumstances these may be unreliable; they might be blocked by solids or gas-bubbles, or the flow profile might be uneven so that the pressures at the tappings are higher or lower than the average on those planes. On these situations multiple tappings can be used, arranged circumferentially around the pipe and joined by a piezometer ring, or (in the case of corner taps) annular slots running completely round the internal circumference of the orifice carrier.

Plate

Standards and handbooks are mainly concerned with *sharp-edged thin* plates. On these, the leading edge is sharp and free of burrs and the cylindrical section of the orifice is short, either because the entire plate is thin or because the downstream edge of the plate is bevelled. Exceptions include the *quarter-circle* or *quadrant-edge* orifice, which has a fully rounded leading edge and no cylindrical section, and the *conical inlet* or *conical entrance* plate which has a bevelled leading edge and a very short cylindrical section. The orifices are normally concentric with the pipe (the *eccentric* orifice is a specific exception) and circular (except on the specific case of the *segmental* or *chord* orifice, on which the plate obstructs just a segment of the pipe). Standards and handbooks stipulate that the upstream surface of the plate is particularly flat and smooth. Sometimes a small drain or vent hole is drilled through the plate where it meets the pipe, to allow condensate or gas bubbles to pass along the pipe.

Pipe

Standards and handbooks stipulate a well-developed flow profile; velocities will be lower at the pipe wall than on the centre but not eccentric or jetting. Similarly the flow downstream of the plate must be unobstructed, otherwise the downstream pressure will be affected. To achieve this, the pipe must be acceptably circular, smooth and straight for stipulated distances. Sometimes when it is impossible to provide enough straight pipe, flow conditioners such as tube bundles or plates with multiple holes are inserted into the pipe to straighten and develop the flow profile, but even these require a further length of straight pipe before the orifice itself. Some standards and handbooks also provide for flows from or into large spaces rather than pipes, stipulating that the region before or after the plate is free of obstruction and abnormalities on the flow.

Theory

Incompressible Flow

By assuming steady-state, incompressible (constant fluid density), inviscid, laminar flow on a horizontal pipe (no change on elevation) with negligible frictional losses, Bernoulli's equation reduces to an equation relating the conservation of energy between two points on the same streamline:

$$p_1 + \frac{1}{2} \cdot \rho \cdot V_2'^2 = p_2 + \frac{1}{2} \cdot \rho \cdot V_2'^2$$

or:

$$p_1 - p_2 = \frac{1}{2} \cdot \rho \cdot V_2'^2 - \frac{1}{2} \cdot \rho \cdot V'^2$$

By continuity equation:

$$q'_v = A_1 \cdot V_1' = A_2 \cdot V_2' \text{ or } V_1' = q'_v / A_1 \text{ and } V_2' = q'_v / A_2$$

$$p_1 - p_2 = \frac{1}{2} \cdot \rho \cdot \left(\frac{q'_v}{A_2}\right)^2 - \frac{1}{2} \cdot \rho \cdot \left(\frac{q'_v}{A_1}\right)^2$$

Solving for q'_v:

$$q'_v = A_2 \sqrt{\frac{2(p_1 - p_2)/\rho}{1 - (A_2 / A_1)^2}}$$

and:

$$q'_v = A_2 \sqrt{\frac{1}{1 - (d/D)^4}} \sqrt{2(p_1 - p_2)/\rho}$$

The above expression for q'_v gives the theoretical volume flow rate. Introducing the beta factor $\beta = d/D$ as well as the discharge coefficient C_d:

$$q_v = C_d A_2 \sqrt{\frac{1}{1 - \beta^4}} \sqrt{2(p_1 - p_2)/\rho}$$

And finally introducing the meter coefficient C which is defined as $C = \dfrac{C_d}{\sqrt{1 - \beta^4}}$ to obtain the final

equation for the volumetric flow of the fluid through the orifice which accounts for irreversible losses:

$$q_v = C A_2 \sqrt{2(p_1 - p_2)/\rho}$$

Multiplying by the density of the fluid to obtain the equation for the mass flow rate at any section on the pipe:

$$q_m = \rho q_v = C A_2 \sqrt{2\rho(p_1 - p_2)}$$

q_v = volumetric flow rate (at any cross-section), m³/s.

q'_v = theoretical volumetric flow rate (at any cross-section), m³/s.

q_m = mass flow rate (at any cross-section), kg/s.

q'_m = theoretical mass flow rate (at any cross-section), kg/s.

C_d = coefficient of discharge, dimensionless.

C = orifice flow coefficient, dimensionless.

A_1 = cross-sectional area of the pipe, m².

A_2 = cross-sectional area of the orifice hole, m².

D = diameter of the pipe, m.

d = diameter of the orifice hole, m.

β = ratio of orifice hole diameter to pipe diameter, dimensionless.

V_1' = theoretical upstream fluid velocity, m/s.

V_2' = theoretical fluid velocity through the orifice hole, m/s.

p_1 = fluid upstream pressure, Pa with dimensions of kg/(m·s²).

p_2 = fluid downstream pressure, Pa with dimensions of kg/(m·s²).

ρ = fluid density, kg/m³.

Deriving the above equations used the cross-section of the orifice opening and is not as realistic as using the minimum cross-section at the vena contracta. On addition, frictional losses may not be negligible and viscosity and turbulence effects may be present. For that reason, the coefficient of discharge C_d is introduced. Methods exist for determining the coefficient of discharge as a function of the Reynolds number.

The parameter $\dfrac{1}{\sqrt{1-\beta^4}}$ is often referred to as the *velocity of approach factor* and multiplying the coefficient of discharge by that parameter (as was done above) produces the flow coefficient C. Methods also exist for determining the flow coefficient as a function of the beta function   and the location of the downstream pressure sensing tap. For rough approximations, the flow coefficient may be assumed to be between 0.60 and 0.75. For a first approximation, a flow coefficient of 0.62 can be used as this approximates to fully developed flow.

An orifice only works well when supplied with a fully developed flow profile. This is achieved by a long upstream length (20 to 40 pipe diameters, depending on Reynolds number) or the use of a flow conditioner. Orifice plates are small and inexpensive but do not recover the pressure drop as well as a venturi, nozzle, or venturi-nozzle does. Venturis also require much less straight pipe upstream. A venturi meter is more efficient, but usually more expensive and less accurate (unless calibrated on a laboratory) than an orifice plate.

Compressible Flow

In general, equation $q_m = \rho\, q_v = C\, A_2 \sqrt{2\,\rho\,(p_1 - p_2)}$ is applicable only for incompressible flows. It can be modified by introducing the expansibility factor, (also called the expansion factor) ε to account for the compressibility of gasses.

$$q_m = \rho_1\, q_{v,1} = C\, \epsilon\, A_2 \sqrt{2\,\rho_1\,(p_1 - p_2)}$$

ε is 1.0 for incompressible fluids and it can be calculated for compressible gases using empirically determined formulae

For smaller values of β (such as restriction plates with β less than 0.25 and discharge from tanks), if the fluid is compressible, the rate of flow depends on whether the flow has become choked. If it is, then the flow may be calculated as shown at choked flow (although the flow of real gases through thin-plate orifices never becomes fully choked). By using a mechanical energy balance, compressible fluid flow on un-choked conditions may be calculated as:

$$q_m = C\,A_2\sqrt{2\,\rho_1\,p_1\left(\frac{\gamma}{\gamma-1}\right)\left[(p_2/p_1)^{2/\gamma}-(p_2/p_1)^{(\gamma+1)/\gamma}\right]}$$

Or

$$q_v = C\,A_2\sqrt{2\frac{p_1}{\rho_1}\left(\frac{\gamma}{\gamma-1}\right)\left[(p_2/p_1)^{2/\gamma}-(p_2/p_1)^{(\gamma+1)/\gamma}\right]}$$

Under choked flow conditions, the fluid flow rate becomes:

$$q_m = C\,A_2\sqrt{\gamma\,\rho_1\,p_1\left(\frac{2}{\gamma+1}\right)^{\frac{\gamma+1}{\gamma-1}}}$$

Or

$$q_v = C\,A_2\sqrt{\gamma\frac{p_1}{\rho_1}\left(\frac{2}{\gamma+1}\right)^{\frac{\gamma+1}{\gamma-1}}}$$

where:

γ = heat capacity ratio $\left(c_p/c_v\right)$, dimensionless ($\gamma \approx 1.4$ for air).

q_m, q_v = mass and volumetric flow rate, respectively, kg/s and m³/s.

ρ_1 = real gas density under upstream conditions, kg/m³.

Computation according to ISO 5167

Flow rates through an orifice plate can be calculated without specifically calibrating the individual flowmeter so long as the construction and installation of the device complies with the stipulations of the relevant standard or handbook. The calculation takes account of the fluid and fluid conditions, the pipe size, the orifice size and the measured differential pressure; it also takes account of the coefficient of discharge of the orifice plate, which depends upon the orifice type and the positions of the pressure tappings. With local pressure tappings (corner, flange and D+D/2), sharp-edged orifices have coefficients around 0.6 to 0.63, while the coefficients for conical entrance plates are on the range 0.73 to 0.734 and for quarter-circle plates 0.77 to 0.85. The coefficients of sharp-edged orifices vary more with fluids and flow rates than the coefficients of conical-entrance and quarter-circle plates, especially at low flows and high viscosities.

For compressible flows such as flows of gases or steam, an *expansibility factor* or *expansion factor* is also calculated. This factor is primarily a function of the ratio of the measured differential pressure to the fluid pressure and so can vary significantly as the flow rate varies, especially at high differential pressures and low static pressures.

The equations provided on American and European national and industry standards and the various coefficients used to differ from each other even to the extent of using different combinations of correction factors, but many are now closely aligned and give identical results; on particular, they use the same *Reader-Harris/Gallagher* equation for the coefficient of discharge for sharp-edged orifice plates. The equations below largely follow the notation of the international standard ISO 5167 and use SI units.

Volume flow rate:

$$q_v = \frac{q_m}{\rho_1}$$

Mass flow rate:

$$q_m = \frac{C}{\sqrt{1-\beta^4}} \epsilon \frac{\pi}{4} d^2 \sqrt{2\,\rho_1 \Delta p}$$

Coefficient of Discharge

Coefficient of discharge for sharp-edged orifice plates with corner, flange or D and D/2 tappings and no drain or vent hole (Reader-Harris/Gallagher equation):

$$C = 0.5961 + 0.0261\beta^2 - 0.216\beta^8 + 0.000521\left(\frac{10^6 \beta}{Re_D}\right)^{0.7} + (0.0188 + 0.0063A)\beta^{3.5}\left(\frac{10^6}{Re_D}\right)^{0.3}$$

$$+ (0.043 + 0.080\exp(-10L_1) - 0.123\exp(-7L_1))(1-0.11A)\frac{\beta^4}{1-\beta^4} - 0.031(M_2' - 0.8M_2'^{1.1})\beta^{1.3}$$

and if D < 71.2mm on which case this further term is added to C:

$$+0011(0.75 - \beta)\left(2.8 - \frac{D}{0.0254}\right)$$

In the equation for C,

$$A = \left(\frac{19000\beta}{Re_D}\right)^{0.8}$$

$$M_2' = \frac{2L_2'}{1-\beta}$$

and only the three following pairs of values for L$_1$ and L'$_2$ are valid:

corner tappings: $L_1 = L_2' = 0$

flange tappings: $L_1 = L_2' = \frac{0.0254}{D}$

D and D/2 tappings:

$$L_1 = 1$$
$$L'_2 = 0.47$$

Expansibility Factor

Expansibility factor, also called expansion factor, for sharp-edged orifice plates with corner, flange or D and D/2 tappings:

If $p_2 / p_1 > 0.75$:5.3.2.2 (at least - standards vary)

$$\epsilon = 1 - (0351 + 0.256\beta^4 + 0.93\beta^8)\left[1 - \left(\frac{p_2}{p_1}\right)^{\frac{1}{\kappa}}\right]$$

but for incompressible fluids, including most liquids

$$\epsilon = 1$$

Where

C = coefficient of discharge, dimensionless

d = internal orifice diameter under operating conditions, m

D = internal pipe diameter under operating conditions, m

p_1 = fluid absolute static pressure on plane of upstream tapping, Pa

p_2 = fluid absolute static pressure on plane of downstream tapping, Pa

q_m = mass flow rate, kg/s

q_v = volume flow rate, m³/s

Re_D = pipe Reynolds number, $\dfrac{4q_m}{\pi\mu D}$, dimensionless

β = diameter ratio of orifice diameter to pipe diameter, $\dfrac{d}{D}$, dimensionless

Δp = differential pressure, Pa

ϵ = expansibility factor, also called expansion factor, dimensionless

κ = isentropic exponent, often approximated by specific heat ratio, dimensionless

μ = dynamic viscosity of the fluid, Pa.s

ρ_1 = fluid density on plane of upstream tapping, kg/m³

Overall Pressure Loss

The overall pressure loss caused by an orifice plate is less than the differential pressure measured across tappings near the plate. For sharp-edged plates such as corner, flange or D and D/2 tappings, it can be approximated by the equation

$$\frac{\Delta\bar{\omega}}{\Delta p} = 1 - \beta^{1.9}$$

Or

$$\frac{\Delta\bar{\omega}}{\Delta p} = \frac{\sqrt{1 - \beta^4(1 - C^2)} - C\beta^2}{\sqrt{1 - \beta^4(1 - C^2)} + C\beta^2}$$

where

$\Delta\bar{\omega}$ = overall pressure loss, Pa.

Flow Nozzle

The flow nozzles are used for flow measurements at high fluid velocities and are more rugged and more resistant to erosion than the sharp edged orifice plate.

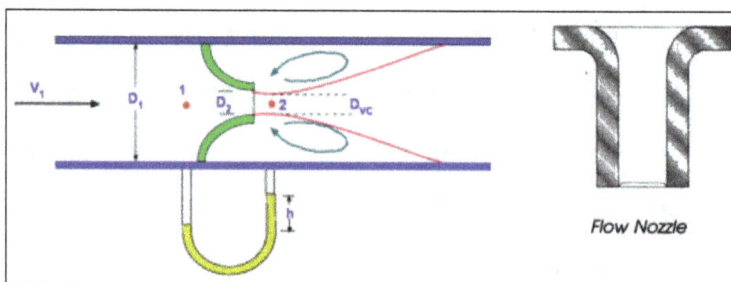

Flow Nozzle

Basically, there are two types of flow nozzles, the long-radius flow nozzles and the I.S.A. (International Federation of the National Standardizing Associations) flow nozzle.

A flow nozzle consists of a convergent inlet whose shape is a quarter ellipse, and a cylindrical throat.

Differential pressure measurement taps are normally located one pipe diameter upstream and one-half diameter downstream from the inlet faces of the nozzle.

For a given diameter and a given differential pressure, it allows measurement of flow rates almost 65% more than that of the orifice plate.

Flow nozzles are manufactured commonly from materials such as stainless steel or chrome-moly steel.

They are made commercially on various configurations, viz. flange type, holding ring type, weld-in type, and throat type.

Flow nozzles should be used at Reynolds numbers of 50,000 or above.

However, data is available for Reynolds number down to 6,000; so it is possible to use nozzles with more viscous fluids.

Flow nozzles have very high coefficients of discharge, typically 0.99 or greater, and a wide range of beta ratios of 0.2 to 0.8.

Advantages

- Its permanent pressure loss is lower than that for an orifice plate.

- It is available on numerous materials.

- It is useful for fluids containing solids that settle.

- It is widely accepted for high-pressure and temperature steam flow.

Disadvantages

- Its cost is higher than orifice plate.

- It is limited to moderate pipe sizes.

- It requires more maintenance (it is necessary to remove a section of pipe to inspect or install it).

Differential Pressure Flow Meter

Differential type flow meter is the most widely used flow measuring device for low viscous fluid on a pipeline that requires an accurate measurement at a reasonable cost. These types of flow meters are generally simple, reliable and offer more flexibility than other flow measurement methods. The most common differential pressure flow meters included Venturi, Orifice, Flow Nozzle, Pitot tube, Dall type.

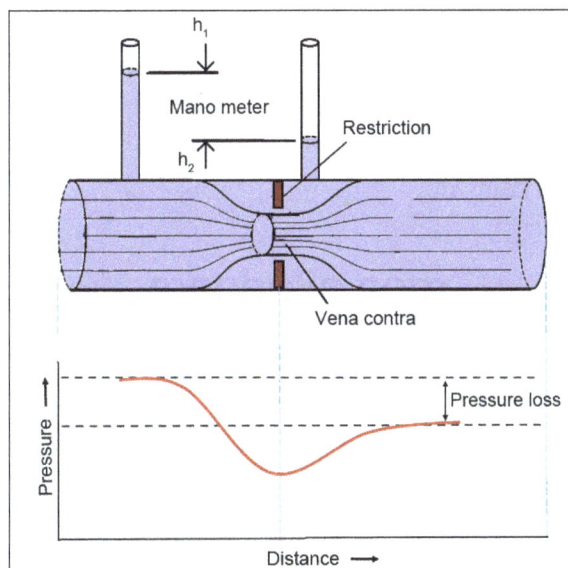

In differential pressure flow meters, a restriction is introduced into the pipe, which results on a pressure drop across the constraint. When such a restriction is placed inside the tube, the velocity of a fluid increases after the constraint, hence the pressure after restriction decreases. The developed differential pressure or head is measured, which provides an indication of flow rate. So the differential pressure flow meters also called head meters. This is shown on above figure.

Vena contracta – Smallest cross-sectional area of the flow stream. At vena contra velocity is maximum, and pressure is minimum.

Differential pressure flowmeter relies on Bernoulli's principle. The relation between flow rate and pressure drop is given by

$$Q = K\sqrt{\Delta P}$$

Q = Flow rate

K = Constant for pipe and fluid type

ΔP = Pressure drop

Primary and Secondary Devices (Element) on Differential Type Flow Meter

These types of flow meter always consist of two components, Primary devices, and Secondary devices. The primary device is placed on the pipe to restrict the flow and hence to develop a differential pressure. They are an orifice plate, venturi, flow nozzle, dall tube, pitot tubes, annubar tube, elbow tap, flume, etc.

The secondary device measures the differential pressure and provides a readout or signal for transmission to the control system. They are manometers, bellow meters, force balance meters, ring balance meters, etc.

With restriction flow meters, calibration of the primary measuring device is not required. The primary device can be selected for compatibility with the specific fluid or application, and the secondary device can be selected for the type or readout or transmission signal desired.

Advantages and Application of Differential Pressure Flow Meter

- Simple and robust construction.
- Easy installation and removal.
- Inexpensive.
- They are adaptable to different flow rate and pipe sizes.
- They can configure on the bidirectional flow.
- Suitable for measuring the flow rate of both liquid and gas.
- Reliable, this device is maintenance–free (very low maintenance) as it has no moving parts.
- It is available on a wide range of sizes.

Disadvantages and Limitation of Differential Pressure Flow Meter

- Causes relatively high permanent pressure drop.

- Flow rate rangeability is low because of the square root relationship between flow rate and pressure head.

- Difficult to measure the flow rate of pulsating flow.

- It may cause condensation and freezing at the connecting piping.

- It cannot measure low flow rate.

- Accuracy depends on the pressure sensor.

- Accuracy depends on many other fluid characteristics such as temperature, compressibility, specific gravity, etc.

- Difficult to use for flow measurement of slurries.

Turbine Flowmeter

A turbine flow meter is a volume sensing device. As liquid or gas passes through the turbine housing, it causes the freely suspended turbine blades to rotate. The velocity of the turbine rotor is directly proportional to the velocity of the fluid passing through the flow meter.

The external pickoff mounted on the body of the flow meter, senses each rotor blade passing, causing the sensor to generate a frequency output. The frequency is directly proportional to the volume of the liquid or gas.

Either a magnetic or modulated carrier (RF) pickup can be used to sense the rotational speed of the turbine rotor.

Depending on your flow meter application, there are many types of turbine flow meters to choose from. And after understanding the application several factors come into effect when choosing a flow meter, such as:

- Fluid Type.

- Viscosity.

- Connection.

- Pipe Sizing.

- Process Temperature (min & max).

- Flow Range (min & max).

- Pressure Range (min & max).

- Accuracy Range.

- Specific Application.

If you need volumetric total flow and/or flow rate measurement, a turbine flow meter is the ideal device. Turbine flow meters are used on a wide variety of liquid and gas flow sensing applications. They can be built to endure high pressure, and high and low temperatures. They offer a high turndown with minimum uncertainty and excellent repeatability. Turbine flowmeters are also simple to install and maintain only requiring periodic recalibration and service.

Turbine Measuring Principle

A turbine flow meter is constructed with rotor and blades that use the mechanical energy of the fluid to rotate the rotor on the flow stream. Blades on the rotor are angled to transform energy from the flow stream into rotational energy. The rotor shaft spins on bearings: when the fluid moves faster, the rotor spins proportionally faster. Shaft rotation can be sensed mechanically or by detecting the movement of the rotor blades.

Rotor movement is often detected magnetically, where movement of the rotor generates a pulse. When the fluid moves faster, more pulses are generated. Turbine flow meter sensors detecting the pulse are typically located external to the flowing stream to avoid material of construction constraints that would result if wetted sensors were used. The RPM of the turbine wheel is directly proportional to the mean flow velocity within the tube diameter and corresponds to the volume flow over a wide range.

A flow transmitter processes the pulse signal to determine the flow of the fluid. Flow transmitter and sensing systems are available to sense flow on both the forward and reverse flow directions. High accuracy turbine flowmeters are available for custody transfer of hydrocarbons and natural gas. This fuel flow meter often incorporates the functionality of a flow computer to correct for pressure, temperature, and fluid properties on order to achieve the desired accuracy for the custody transfer application.

Care should be taken when using a turbine flow meter on fluids that are non-lubricating because the flowmeter can become inaccurate and fail if its bearings prematurely wear. A turbine flow meter can be outfitted with grease fittings for applications with non-lubricating fluids. On addition, a turbine flow meter that is designed for a specific purpose, such as natural gas service, can often operate over a limited range of temperatures (such as up to 140 °F or 60 °C) whereby operation at higher temperatures can damage the flowmeter.

A turbine flow meter is less accurate at low flow rates due to rotor/bearing drag that slows the rotor. Care should be taken when operating these flowmeters above approximately 5 percent of maximum flow. A turbine flow meter should not be operated at high velocity because premature bearing wear and/or damage can occur. When measuring fluids that are non-lubricating, bearing wear can cause the flowmeter to become inaccurate and fail. Bearing replacement may be needed on some applications to maintain good accuracy. Application on dirty fluids should generally be avoided so as to reduce the possibility of flowmeter wear and bearing damage.

Turbine flow meters have moving parts that are subject to degradation with time and use. Abrupt transitions from gas flowmeter applications to liquid flowmeter use should be avoided because they can mechanically stress the flowmeter, degrade accuracy, and/or damage the flow meter. These conditions generally occur when filling the pipe and under slug flow conditions. Using the turbine flow meter for two-phase flow conditions such as steam flow metering applications can also cause a turbine flow meter to measure inaccurately.

A turbine flow meter measures the velocity of liquids, gases and vapors on pipes, such as hydrocarbons on fuel flow measurement, chemical flow metering, water flow metering, cryogenic liquid flow metering, air or gas flow metering, and general industrial flow metering. High accuracy turbine flowmeters are available for custody transfer of hydrocarbons and natural gas. A mass flow computer is often used on custody-transfer applications to correct for pressure, temperature and fluid properties on order to achieve the desired accuracy. Other low viscosity applications are tap and demineralized water, fuel flow meter solvents, and pharmaceutical fluids.

Ultrasonic Flow Meter

An ultrasonic flow meter is a type of flow meter that measures the velocity of a fluid with ultrasound to calculate volume flow. Using ultrasonic transducers, the flow meter can measure the average velocity along the path of an emitted beam of ultrasound, by averaging the difference on measured transit time between the pulses of ultrasound propagating into and against the direction of the flow or by measuring the frequency shift from the Doppler effect. Ultrasonic flow meters are affected by the acoustic properties of the fluid and can be impacted by temperature, density, viscosity and suspended particulates depending on the exact flow meter. They vary greatly on purchase price but are often inexpensive to use and maintain because they do not use moving parts, unlike mechanical flow meters.

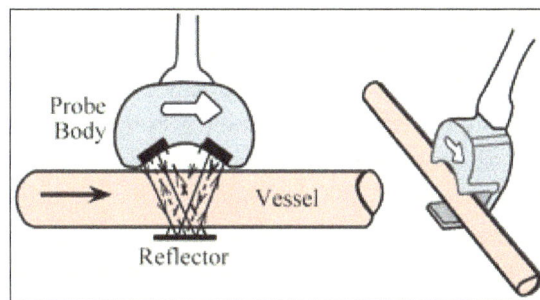

Schematic view of a flow sensor.

Means of Operation

There are three different types of ultrasonic flow meters. Transmission (or contrapropagating transit-time) flow meters can be distinguished into in-line (intrusive, wetted) and clamp-on (non-intrusive) varieties. Ultrasonic flow meters that use the Doppler shift are called reflection or Doppler flow meters. The third type is the open-channel flow meter.

Principle

Time Transit Flow Meter

Ultrasonic flow meters measure the difference between the transit time of ultrasonic pulses propagating with and against the flow direction. This time difference is a measure for the average velocity of the fluid along the path of the ultrasonic beam. By using the absolute transit times t_{up} and t_{down}, both the averaged fluid velocity v and the speed of sound c can be calculated. Using these two transit times, the distance between receiving and transmitting transducers L and the inclination angle α, if we assume that sound has to go against the flow when going up and along the flow when returning down, then one can write the following equations from the definition of velocity:

$$c - v\cos\alpha = \frac{L}{t_{up}} \quad and \quad c + v\cos\alpha = \frac{L}{t_{down}}$$

By adding and subtracting the above equations we get,

$$v = \frac{L}{2\cos(\alpha)} \frac{t_{up} - t_{down}}{t_{up}\, t_{down}} \quad and \quad c = \frac{L}{2} \frac{t_{up} + t_{down}}{t_{up}\, t_{down}}$$

where v is the average velocity of the fluid along the sound path and c is the speed of sound.

Doppler Shift Flow Meters

Another method on ultrasonic flow metering is the use of the Doppler shift that results from the reflection of an ultrasonic beam off sonically reflective materials, such as solid particles or entrained air bubbles on a flowing fluid, or the turbulence of the fluid itself, if the liquid is clean.

Doppler flowmeters are used for slurries, liquids with bubbles, gases with sound-reflecting particles.

This type of flow meter can also be used to measure the rate of blood flow, by passing an ultrasonic beam through the tissues, bouncing it off a reflective plate, then reversing the direction of the beam and repeating the measurement, the volume of blood flow can be estimated. The frequency of the transmitted beam is affected by the movement of blood on the vessel and by comparing the frequency of the upstream beam versus downstream the flow of blood through the vessel can be measured. The difference between the two frequencies is a measure of true volume flow. A wide-beam sensor can also be used to measure flow independent of the cross-sectional area of the blood vessel.

Open Channel Flow Meters

In this case, the ultrasonic element is actually measuring the height of the water on the open channel; based on the geometry of the channel, the flow can be determined from the height. The ultrasonic sensor usually also has a temperature sensor with it because the speed of sound on air is affected by the temperature.

Thermal Mass Flow Meter

Thermal mass flow meters measure the mass flowrate of gases and liquids directly. Volumetric measurements are affected by all ambient and process conditions that influence unit volume or indirectly affect pressure drop, while mass flow measurement is unaffected by changes on viscosity, density, temperature, or pressure. Thermal mass flow meters are often used on monitoring or controlling mass-related processes such as chemical reactions that depend on the relative masses of unreacted ingredients. On detecting the mass flow of compressible vapors and gases, the measurement is unaffected by changes on pressure and/or temperature. One of the capabilities of thermal mass flow meters is to accurately measure low gas flowrates or low gas velocities (under 25 ft. per minute)--much lower than can be detected with any other device.

Thermal flow meters provide high rangeability (10:1 to 100:1) if they are operated on constant-temperature-difference mode. On the other hand, if heat input is constant, the ability to detect very small temperature differences is limited, and both precision and rangeability drop off. At normal flows, measurement errors are usually on the 1-2% full-scale range.

This meter is available on high-pressure and high-temperature designs, and on special materials including glass, Monel®, and PFA. Flow-through designs are used to measure small flows of pure substances (heat capacity is constant if a gas is pure), while bypass and probe-type designs can detect large flows on ducts, flare stacks, and dryers.

Theory of Operation

Thermal mass flow meters are most often used for the regulation of low gas flows. They operate either by introducing a known amount of heat into the flowing stream and measuring an associated temperature change or by maintaining a probe at a constant temperature and measuring the energy required to do so. The components of a basic thermal mass flow meter include two temperature sensors and an electric heater between them. The heater can protrude into the fluid stream or can be external to the pipe.

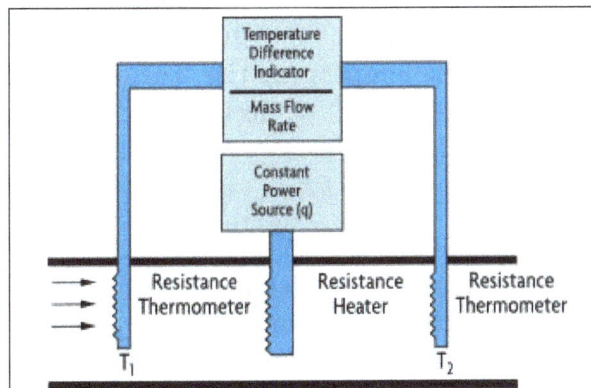

Immersion Heater.

In the direct-heat version, a fixed amount of heat (q) is added by an electric heater. As the process fluid flows through the pipe, resistance temperature detectors (RTDs) measure the temperature rise, while the amount of electric heat introduced is held constant.

The mass flow (m) is calculated on the basis of the measured temperature difference (T2 - T1), the meter coefficient (K), the electric heat rate (q), and the specific heat of the fluid (Cp), as follows:

$$m = Kq/(C_p(T_2 - T_1))$$

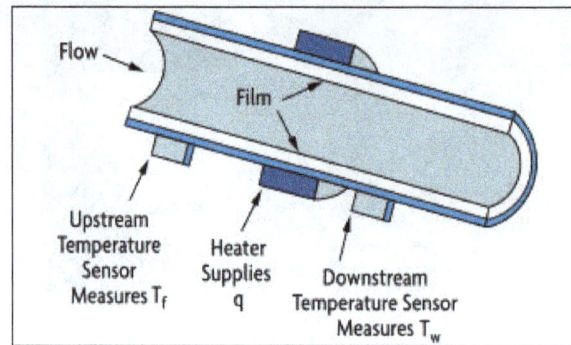

Externally-Heated Tube.

Heated Tube Design

Heated-tube flow meters were developed to protect the heater and sensor elements from corrosion and any coating effects of the process. By mounting the sensors externally to the piping (Figure 5-8B), the sensing elements respond more slowly, and the relationship between mass flow and temperature difference becomes nonlinear. This nonlinearity results from the fact that the heat introduced is distributed over some portion of the pipe's surface and transferred to the process fluid at different rates along the length of the pipe.

The pipe wall temperature is highest near the heater (detected as Tw on Figure 5-8B), while, some distance away, there is no difference between wall and fluid temperature. Therefore, the temperature of the unheated fluid (Tf) can be detected by measuring the wall temperature at this location further away from the heater. This heat transfer process is non-linear, and the corresponding equation differs from the one above as follows:

$$m^{0.8} = Kq/(C_p(T_w - T_f))$$

This flow meter has two operating modes: one measures the mass flow by keeping the electric power input constant and detecting the temperature rise. The other mode holds the temperature difference constant and measures the amount of electricity needed to maintain it. This second mode of operation provides for a much higher meter rangeability.

Bypass-type Design

The bypass version of the thermal mass flow meter was developed to measure larger flow rates. It consists of a thin-walled capillary tube (approximately 0.125 on diameter) and two externally wound self-heating resistance temperature detectors (RTDs) that both heat the tube and measure the resulting temperature rise. The meter is placed on a bypass around a restriction on the main pipe and is sized to operate on the laminar flow region over its full operating range.

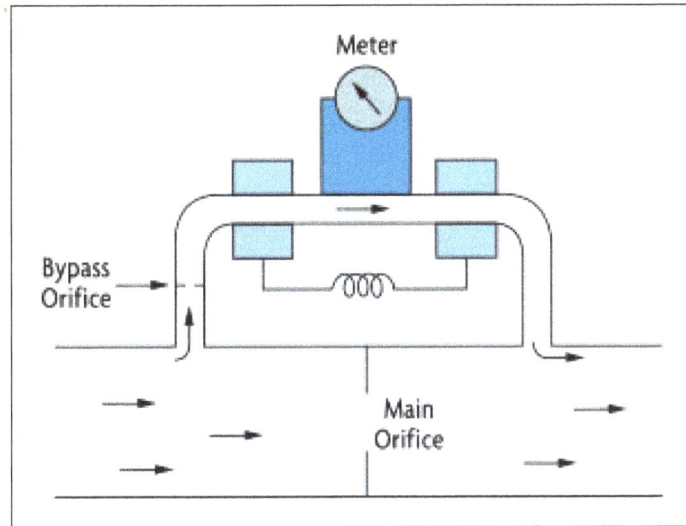

Bypass Uses Small Percent of Stream.

When there is no flow, the heaters raise the bypass-tube temperature to approximately 160°F above ambient temperature. Under this condition, a symmetrical temperature distribution exists along the length of the tube. When flow is taking place, the gas molecules carry the heat downstream, and the temperature profile is shifted on the direction of the flow. A Wheatstone bridge connected to the sensor terminals converts the electrical signal into a mass flow rate proportional to the change on temperature.

The small size of the bypass tube makes it possible to minimize electric power consumption and to increase the speed of response of the measurement. On the other hand, because of the small size, filters are necessary to prevent plugging. One serious limitation is the high-pressure drop (up to 45 psi) needed to develop laminar flow. This is typically acceptable only for high-pressure gas applications where the pressure needs to be reduced on any case.

This is a low accuracy (2% full scale), low maintenance, and low-cost flow meter. Electronic packages within the units allow for data acquisition, chart recording, and computer interfacing. These devices are popular on the semiconductor processing industry. Modern day units are also available as complete control loops, including a controller and automatic control valve.

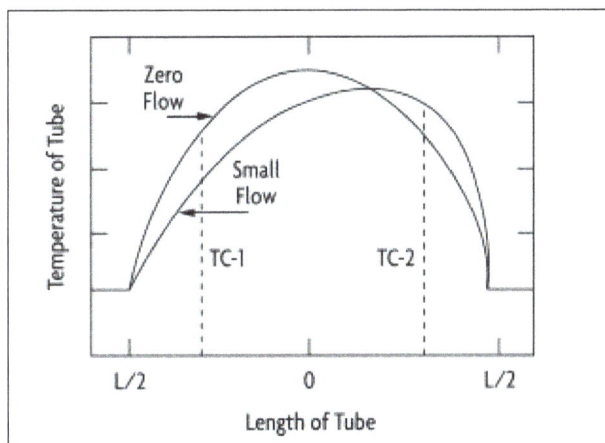

Temperature Profile.

Air Velocity Probes

Probe-style mass flow meters are used to measure air flows and are insensitive to the presence of moderate amounts of dust. They maintain a temperature differential between two RTDs mounted on the sensor tube. The upper sensor measures the ambient temperature of the gas and continuously maintains the second RTD (near the tip of the probe) at 60°F above ambient. The higher the gas velocity, the more current is required to maintain the temperature differential.

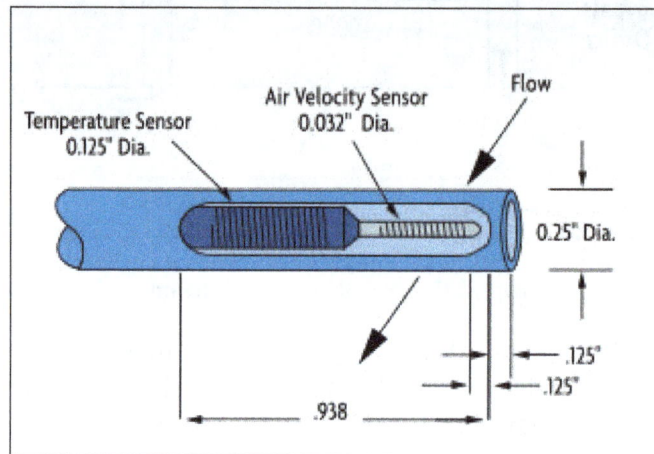

Probe Configuration.

Another version of the velocity probe is the venturi-type thermal mass flow meter, which places a heated mass flow sensor at the minimum diameter of a venturi flow element and a temperature compensation probe downstream. An inlet screen mixes the flow to make the temperature uniform. This design is used for both gas and liquid measurement (including slurries), with flow range a function of the size of the venturi. Pressure drop is relatively low, and precision is dependent upon finding the proper probe insertion depth.

A flow switch version is also available that contains two temperature sensors on the tip. One of the sensors is heated and the temperature difference is a measure of velocity. The switch can be used to detect high or low flow within 5%.

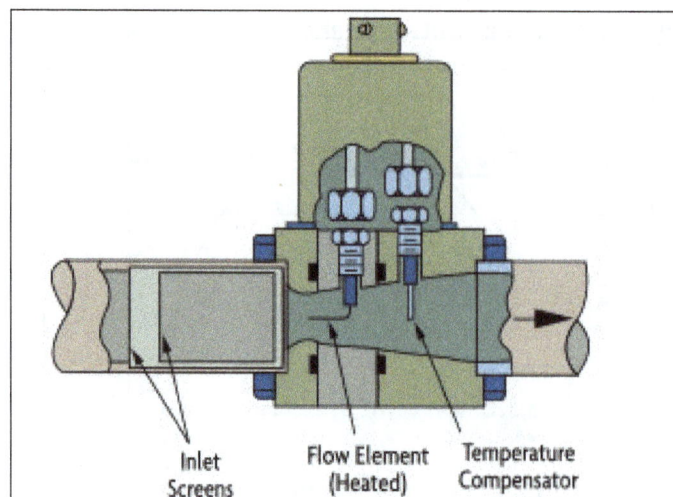

Venturi Insertion.

Hot-wire Anemometers

A hot-wire anemometer consists of an electrically heated, fine-wire element (0.00016 inch on diameter and 0.05 inch long) supported by needles at its ends. Tungsten is used as the wire material because of its strength and high-temperature coefficient of resistance. When placed on a moving stream of gas, the wire cools; the rate of cooling corresponds to the mass flowrate.

The circuitry of the heated sensing element is controlled by one of two types of solid-state electronic circuits: constant-temperature or constant-power. The constant-temperature sensor maintains a constant temperature differential between a heated sensor and a reference sensor; the amount of power required to maintain the differential is measured as an indication of the mass flow rate.

Constant-temperature anemometers are popular because of their high-frequency response, low electronic noise level, immunity from sensor burnout when airflow suddenly drops, compatibility with hot-film sensors, and their applicability to liquid or gas flows.

Constant-power anemometers do not have a feedback system. Temperature is simply proportional to flowrate. They are less popular because their zero-flow reading is not stable, temperature and velocity response is slow, and temperature compensation is limited.

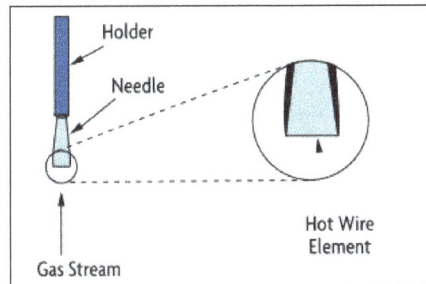

Hot-wire Anemometer.

Air Duct Traversing

Anemometers are widely used for air duct balancing. This is accomplished by placing multiple anemometers on a cross-section of the duct or gas pipe and manually recording the velocity readings at numerous points. The mass flow rate is obtained by calculating the mean velocity and multiplying this by the density and by the cross-sectional area measurement of the duct.

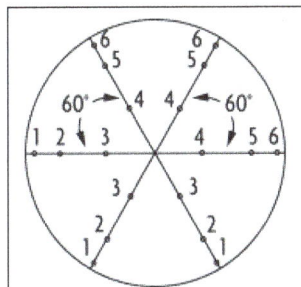

Circular Measuring Station.

For cylindrical ducts, the log-linear method of traversing provides the highest accuracy because it takes into account the effects of friction along the walls of the duct. Because of the number of measurements, air duct traversing is a time-consuming task. Microprocessor- based anemometers are available to automate this procedure.

Because of the small size and fragility of the wire, hot-wire anemometers are susceptible to dirt build-up and breakage. A positive consequence of their small mass is a fast speed of response. They are widely used on HVAC and ventilation applications. Larger and more rugged anemometers are also available for more demanding industrial applications. To ensure the proper formation of the velocity profile, a straight duct section is usually provided upstream of the anemometer station (usually 10 diameters long). A conditioning nozzle is used to eliminate boundary layer effects. If there is no room for the straight pipe section, a honeycomb flow straightener can be incorporated into the sensor assembly.

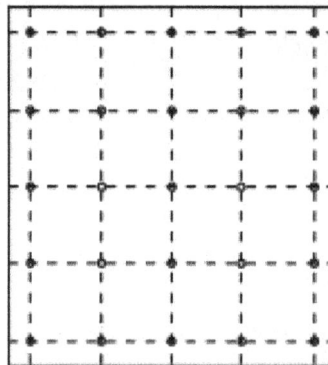

Rectangular Measuring Station.

Uses and Limitations

Thermal mass flow meters can have very high rangeability and reasonable accuracy, but they also have serious limitations. Potential problems include the condensation of moisture (in saturated gases) on the temperature detector. Such condensation will cause the thermometer to read low and can lead to corrosion. Coating or material build-up on the sensor also will inhibit heat transfer and cause the meter to read low. Additional potential sources of error include variations on the specific heat caused by changes on the gas's composition.

Some common gas-flow applications for thermal mass flow meters include combustion air measurement on large boilers, semiconductor process gas measurement, air sampling on nuclear power plants, process gas measurements on the chemical and petrochemical industries, research and development applications, gas chromatography, and filter and leak testing. While hot-wire anemometers are best suited for clean gases at low velocities, venturi meters can also be considered for some liquid (including slurry) flow applications. Thermal mass flow meters are well suited for high rangeability measurements of very low flows, but also can be used on measuring large flows such as combustion air, natural gas, or the distribution of compressed air.

Vortex Flowmeter

Vortex flowmeter is a flowmeter for measuring fluid flow rates on an enclosed conduit.

Composition of Vortex Flowmeter

A vortex flowmeter comprising: a flow sensor operable to sense pressure variations due to vortex-shedding of a fluid on a passage and to convert the pressure variations to a flow sensor signal, on the form of an electrical signal; and a signal processor operable to receive the flow sensor signal and to generate an output signal corresponding to the pressure variations due to vortex-shedding of the fluid on the passage.

Working Principle

When the medium flows through the Bluff body at a certain speed, an alternately arranged vortex belt is generated behind the sides of the Bluff body, called the "von Kármán vortex". Since both sides of the vortex generator alternately generate the vortex, the pressure pulsation is generated on both sides of the generator, which makes the detector produce alternating stress. The piezoelectric element encapsulated on the detection probe body generates an alternating charge signal with the same frequency as the vortex, under the action of alternating stress. The frequency of these pulses is directly proportional to flow rate. The signal is sent to the intelligent flow totalizer to be processed after being amplified by the pre-amplifier.

In certain range of Reynolds number(2×10^4~7×10^6),the relationship among vortex releasing frequency, fluid velocity, and vortex generator facing flow surface width can be expressed by the following equation:

$$f=St\times V/d$$

Wherein, f is the releasing frequency of Carmen vortex, St is the Strouhal number, V is velocity, and d is the width of the triangular cylinder.

Industrial Applications

The vortex flowmeter is a broad-spectrum flow meter which can be used for metering, measurement and control of most steam, gas and liquid flow for a very unique medium versatility, high stability and high reliability with no moving parts, simple structure and low failure rate. The vortex flowmeter is relatively economical because of its simple flow measurement system and ease of maintenance. It is widely used on heavy industrial applications, power facilities, and energy industries, particularly on steam processes.

Rotameter

A rotameter is a device that measures the volumetric flow rate of fluid on a closed tube.

It belongs to a class of meters called variable area meters, which measure flow rate by allowing the cross-sectional area the fluid travels through to vary, causing a measurable effect.

The first variable area meter with rotating float was invented by Karl Kueppers on Aachen on 1908. This is described on the German patent 215225. Felix Meyer founded the company *"Deutsche Rotawerke GmbH"* on Aachen recognizing the fundamental importance of this invention.

They improved this invention with new shapes of the float and of the glass tube. Kueppers invented the special shape for the inside of the glass tube that realized a symmetrical flow scale.

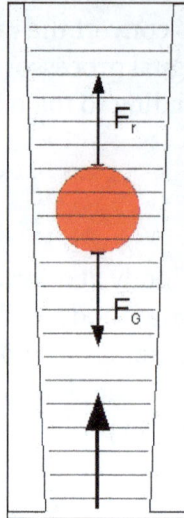

The brand name Rotameter was registered by the British company GEC Elliot automation, Rotameter Co. On many other countries the brand name Rotameter is registered by Rota Yokogawa GmbH & Co. KG on Germany which is now owned by Yokogawa Electric Corp.

TecFluid-CG34-2500 for water flow measurement.

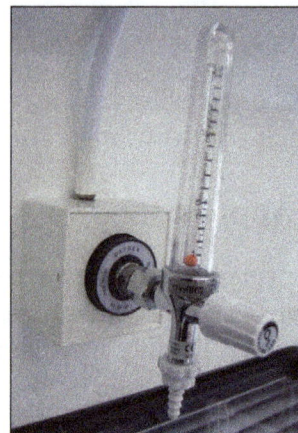

Medical oxygen regulator with rotameter.

Implementation

A rotameter consists of a tapered tube, typically made of glass with a 'float' (a shaped weight, made either of anodized aluminum or a ceramic), inside that is pushed up by the drag force of the flow and pulled down by gravity. The drag force for a given fluid and float cross section is a function of flow speed squared only.

A higher volumetric flow rate through a given area increases flow speed and drag force, so the float will be pushed upwards. However, as the inside of the rotameter is cone shaped (widens), the area around the float through which the medium flows increases, the flow speed and drag force decrease until there is mechanical equilibrium with the float's weight.

Floats are made on many different shapes, with spheres and ellipsoids being the most common. The float may be diagonally grooved and partially colored so that it rotates axially as the fluid passes. This shows if the float is stuck since it will only rotate if it is free. Readings are usually taken at the top of the widest part of the float; the center for an ellipsoid, or the top for a cylinder. Some manufacturers use a different standard.

The "float" must not float on the fluid: it has to have a higher density than the fluid, otherwise it will float to the top even if there is no flow.

The mechanical nature of the measuring principle provides a flow measurement device that does not require any electrical power. If the tube is made of metal, the float position is transferred to an external indicator via a magnetic coupling. This capability has considerably expanded the range of applications for the variable area flowmeter, since the measurement can observed remotely from the process or used for automatic control.

Advantages

- A rotameter requires no external power or fuel, it uses only the inherent properties of the fluid, along with gravity, to measure flow rate.

- A rotameter is also a relatively simple device that can be mass manufactured out of cheap materials, allowing for its widespread use.

- Since the area of the flow passage increases as the float moves up the tube, the scale is approximately linear.

- Clear glass is used which is highly resistant to thermal shock and chemical action.

Disadvantages

- Due to its reliance on the ability of the fluid or gas to displace the float, graduations on a given rotameter will only be accurate for a given substance at a given temperature. The main property of importance is the density of the fluid; however, viscosity may also be significant. Floats are ideally designed to be insensitive to viscosity; however, this is seldom verifiable from manufacturers' specifications. Either separate rotameters for different densities and viscosities may be used, or multiple scales on the same rotameter can be used.

- Because operation of a rotameter depends on the force of gravity for operation, a rotameter must be oriented vertically. Significant error can result if the orientation deviates significantly from the vertical.

- Due to the direct flow indication the resolution is relatively poor compared to other measurement principles. Readout uncertainty gets worse near the bottom of the scale. Oscillations of the float and parallax may further increase the uncertainty of the measurement.

- Since the float must be read through the flowing medium, some fluids may obscure the reading. A transducer may be required for electronically measuring the position of the float.

- Rotameters are not easily adapted for reading by machine; although magnetic floats that drive a follower outside the tube are available.

- Rotameters are not generally manufactured on sizes greater than 6 inches/150 mm, but bypass designs are sometimes used on very large pipes.

Pitot Tube

A pitot tube, also known as pitot probe, is a flow measurement device used to measure fluid flow velocity. The pitot tube was invented by the French engineer Henri Pitot on the early 18th century and was modified to its modern form on the mid-19th century by French scientist Henry Darcy. It is widely used to determine the airspeed of an aircraft, water speed of a boat, and to measure liquid, air and gas flow velocities on certain industrial applications.

Aircraft use pitot tubes to measure airspeed. This example, from an
Airbus A380, combines a pitot tube (right) with a static port and
an angle-of-attack vane (left). Air-flow is right to left.

Theory of Operation

The basic pitot tube consists of a tube pointing directly into the fluid flow. As this tube contains fluid, a pressure can be measured; the moving fluid is brought to rest (stagnates) as there is no outlet to allow flow to continue. This pressure is the stagnation pressure of the fluid, also known as the total pressure or (particularly on aviation) the pitot pressure.

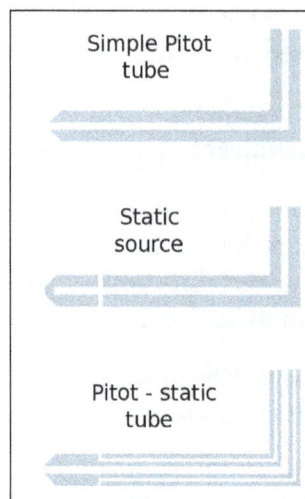

Types of pitot tubes.

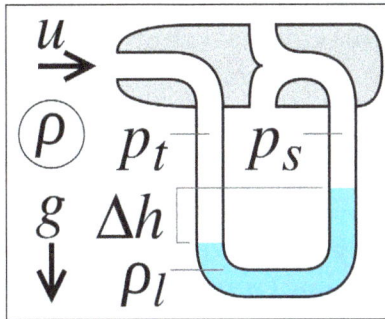

A pitot-static tube connected to a manometer.

Pitot tube on Kamov Ka-26 helicopter.

Pitot tube on a Renault Formula One car.

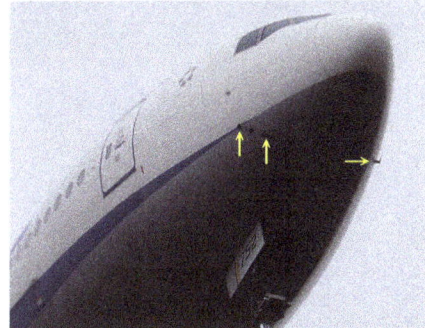

Location of pitot tubes on a Boeing 777.

The measured stagnation pressure cannot itself be used to determine the fluid flow velocity (air-speed on aviation). However, Bernoulli's equation states:

Stagnation Pressure = Static Pressure + Dynamic Pressure

Which can also be written;

$$p_t = p_s + \left(\frac{\rho u^2}{2} \right).$$

Solving that for flow velocity gives;

$$u = \sqrt{\frac{2 \left(p_t - p_s \right)}{\rho}},$$

where,

   u is the flow velocity;

   p_t is the stagnation or total pressure;

   p_s is the static pressure;

   and ρ is the fluid density.

NOTE: The above equation applies only to fluids that can be treated as incompressible. Liquids are treated as incompressible under almost all conditions. Gases under certain conditions can be approximated as incompressible.

The dynamic pressure, then, is the difference between the stagnation pressure and the static pressure. The dynamic pressure is then determined using a diaphragm inside an enclosed container. If the air on one side of the diaphragm is at the static pressure, and the other at the stagnation pressure, then the deflection of the diaphragm is proportional to the dynamic pressure.

In aircraft, the static pressure is generally measured using the static ports on the side of the fuselage. The dynamic pressure measured can be used to determine the indicated airspeed of the aircraft. The diaphragm arrangement described above is typically contained within the airspeed indicator, which converts the dynamic pressure to an airspeed reading by means of mechanical levers.

Instead of separate pitot and static ports, a pitot-static tube (also called a Prandtl tube) may be employed, which has a second tube coaxial with the pitot tube with holes on the sides, outside the direct airflow, to measure the static pressure.

If a liquid column manometer is used to measure the pressure difference $\Delta p \equiv p_t - p_s$;

$$\Delta h = \frac{\Delta p}{\rho_l g},$$

where

> Δh is the height difference of the columns;
>
> ρ_l is the density of the liquid on the manometer;
>
> g is the standard acceleration due to gravity.

Therefore,

$$u = \sqrt{\frac{2\Delta h\, \rho_l g}{\rho}}.$$

Aircraft

A pitot-static system is a system of pressure-sensitive instruments that is most often used on aviation to determine an aircraft's airspeed, Mach number, altitude, and altitude trend. A pitot-static system generally consists of a pitot tube, a static port, and the pitot-static instruments. Errors on pitot-static system readings can be extremely dangerous as the information obtained from the pitot static system, such as airspeed, is potentially safety-critical.

Several commercial airline incidents and accidents have been traced to a failure of the pitot-static system. Examples include Austral Líneas Aéreas Flight 2553, Northwest Airlines Flight 6231, Birgenair Flight 301 and one of the two X-31s. The French air safety authority BEA said that pitot tube icing was a contributing factor on the crash of Air France Flight 447 into the Atlantic Ocean. on 2008 Air Caraïbes reported two incidents of pitot tube icing malfunctions on its A330s.

Birgenair Flight 301 had a fatal pitot tube failure which investigators suspected was due to insects creating a nest inside the pitot tube; the prime suspect is the black and yellow mud dauber wasp.

Aeroperú Flight 603 had a pitot-static system failure due to the cleaning crew leaving the static port blocked with tape.

Industry Applications

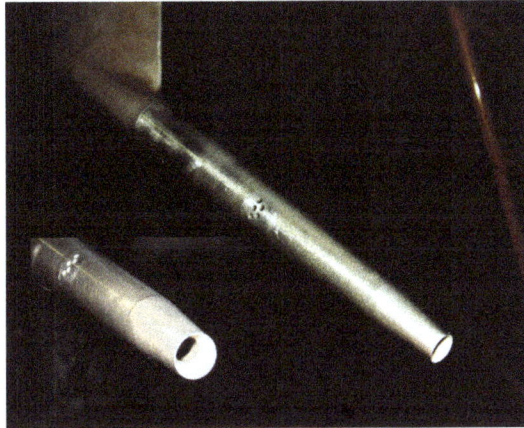

Pitot tube from an F/A-18.

In industry, the flow velocities being measured are often those flowing on ducts and tubing where measurements by an anemometer would be difficult to obtain. On these kinds of measurements, the most practical instrument to use is the pitot tube. The pitot tube can be inserted through a small hole on the duct with the pitot connected to a U-tube water gauge or some other differential pressure gauge for determining the flow velocity inside the ducted wind tunnel. One use of this technique is to determine the volume of air that is being delivered to a conditioned space.

Weather instruments at Mount Washington Observatory. Pitot tube static anemometer is on the right.

The fluid flow rate on a duct can then be estimated from:

Volume flow rate (cubic feet per minute) = duct area (square feet) × flow velocity (feet per minute)

Volume flow rate (cubic meters per second) = duct area (square meters) × flow velocity (meters per second)

In aviation, airspeed is typically measured on knots.

In weather stations with high wind speeds, the pitot tube is modified to create a special type of anemometer called pitot tube static anemometer.

Mass Flow Meter

A mass flow meter, also known as an inertial flow meter is a device that measures mass flow rate of a fluid traveling through a tube. The mass flow rate is the mass of the fluid traveling past a fixed point per unit time.

A mass flow meter of the coriolis type.

The mass flow meter does not measure the volume per unit time (e.g., cubic meters per second) passing through the device; it measures the mass per unit time (e.g., kilograms per second) flowing through the device. Volumetric flow rate is the mass flow rate divided by the fluid density. If the density is constant, then the relationship is simple. If the fluid has varying density, then the relationship is not simple. The density of the fluid may change with temperature, pressure, or composition, for example. The fluid may also be a combination of phases such as a fluid with entrained bubbles. Actual density can be determined due to dependency of sound velocity on the controlled liquid concentration.

Operating Principle of a Coriolis Flow Meter

There are two basic configurations of coriolis flow meter: the curved tube flow meter and the straight tube flow meter.

The animations on the right do not represent an actually existing Coriolis flow meter design. The purpose of the animations is to illustrate the operating principle, and to show the connection with rotation.

Fluid is being pumped through the mass flow meter. When there is mass flow, the tube twists slightly. The arm through which fluid flows away from the axis of rotation must exert a force on the fluid, to increase its angular momentum, so it bends backwards. The arm through which fluid is pushed back to the axis of rotation must exert a force on the fluid to decrease the fluid's angular momentum again, hence that arm will bend forward. On other words, the inlet arm (containing an outwards directed flow), is lagging behind the overall rotation, the part which on rest is parallel to the axis is now skewed, and the outlet arm (containing an inwards directed flow) leads the overall rotation.

The fluid is led through two parallel tubes. An actuator (not shown) induces equal counter vibrations on the sections parallel to the axis, to make the measuring device less sensitive to outside vibrations. The actual frequency of the vibration depends on the size of the mass flow meter, and ranges from 80 to 1000 Hz. The amplitude of the vibration is too small to be seen, but it can be felt by touch.

When no fluid is flowing, the motion of the two tubes is symmetrical, as shown on the left animation. The animation on the right illustrates what happens during mass flow: some twisting of the tubes. The arm carrying the flow away from the axis of rotation must exert a force on the fluid to accelerate the flowing mass to the vibrating speed of the tubes at the outside (increase of absolute angular momentum), so it is lagging behind the overall vibration. The arm through which fluid is pushed back towards the axis of movement must exert a force on the fluid to decrease the fluid's absolute angular speed (angular momentum) again, hence that arm leads the overall vibration.

The inlet arm and the outlet arm vibrate with the same frequency as the overall vibration, but when there is mass flow the two vibrations are out of sync: the inlet arm is behind, the outlet arm is ahead. The two vibrations are shifted on phase with respect to each other, and the degree of phase-shift is a measure for the amount of mass that is flowing through the tubes.

Density and Volume Measurements

The mass flow of a u-shaped coriolis flow meter is given as: $Q_m = \dfrac{K_u - I_u \omega^2}{2Kd^2} \tau$

where K_u is the temperature dependent stiffness of the tube, K a shape-dependent factor, d the width, τ the time lag, ω the vibration frequency and I_u the inertia of the tube. As the inertia of the tube depend on its contents, knowledge of the fluid density is needed for the calculation of an accurate mass flow rate.

If the density changes too often for manual calibration to be sufficient, the coriolis flow meter can be adapted to measure the density as well. The natural vibration frequency of the flow tubes depends on the combined mass of the tube and the fluid contained on it. By setting the tube on motion and measuring the natural frequency, the mass of the fluid contained on the tube can be deduced. Dividing the mass on the known volume of the tube gives us the *density* of the fluid.

An instantaneous density measurement allows the calculation of flow on volume per time by dividing mass flow with density.

Calibration

Both mass flow and density measurements depend on the vibration of the tube. Calibration is affected by changes on the rigidity of the flow tubes.

Changes on temperature and pressure will cause the tube rigidity to change, but these can be compensated for through pressure and temperature zero and span compensation factors.

Additional effects on tube rigidity will cause shifts on the calibration factor over time due to degradation of the flow tubes. These effects include pitting, cracking, coating, erosion or corrosion.

It is not possible to compensate for these changes dynamically, but efforts to monitor the effects may be made through regular meter calibration or verification checks. If a change is deemed to have occurred, but is considered to be acceptable, the offset may be added to the existing calibration factor to ensure continued accurate measurement.

LVDT Flow Meter

An LVDT (Linear Variable Differential Transformer) Variable Area Rotameter, is a meter designed to measure the flow rate of a fluid or gas.

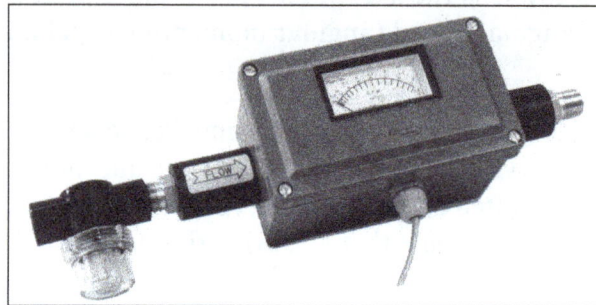

LVDT flow meter.

The flow meter utilizes a unique combination of a tappered metering cone on series with a piston. The position of the metal piston is sensed by the LVDT circuitry and is then translated into a flow rate. This non linear signal can be directly displayed or linearized with an electrical output. Advantages include the ability to externally measure very low flow rates.

Cyclonic Flow Meter

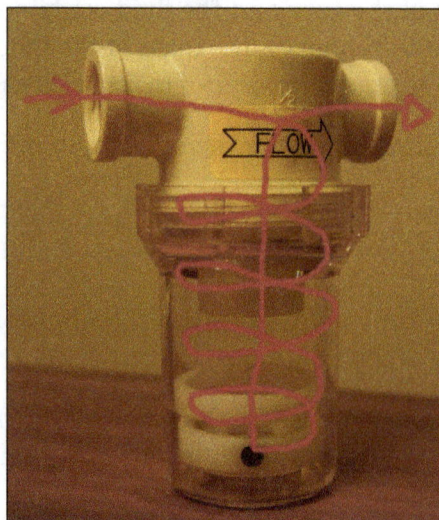

First model of a Cyclonic Flow Meter.

CYCLONIC FLOW CURVE EXAMPLE

K-FACTOR

FORMULA
$K = 60*F/Q$

EXAMPLE
$844 = 60*174/12.37$

ZERO OFFSET
$= .63$

Q=TOTAL GALLONS
F&Q ARE DELTA
VARIABLES

FREQ

GPM

ZERO OFFSET
.63 gpm

(zero offset is subracted from curve total
max gallons to obtain variable "total gallons"
or "Q")

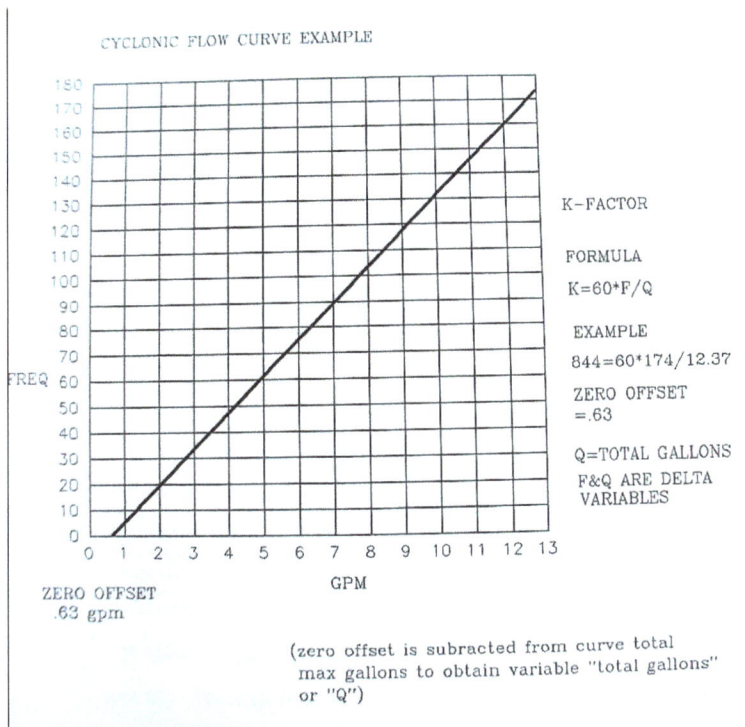

Cyclonic Flow Meter Curve.

A Cyclonic Flow Meter is a meter designed to measure the flow rate of a fluid (or gas) within a cylindrical chamber without a differential pressure directly across the rotating member, and without the use of any bearing (such as on a Turbine meter or Paddle wheel). The ring rotates on a "dead end" of the fluid flow, at the tip of the swirling cyclone. Bypass configuration: A small Cyclonic Flow Meter can be assembled on a "by-pass" configuration across an orifice plate and calibrated as a single unit to obtain high flow rates.

Coriolis Flow Meter

A mass flow meter, also known as inertial flow meter and coriolis flow meter, is a device that measures how much liquid is flowing through a tube. It does not measure the volume of the liquid passing through the tube, it measures the amount of mass flowing through the device.

Volumetric flow rate metering is proportional to mass flow rate only when the density of the fluid is constant. If the fluid has varying density, or contains bubbles, then the volume flow rate multiplied by the density is not an accurate measure of the mass flow rate.

In a mass flow meter the fluid is contained on a smooth tube, with no moving parts that would need to be cleaned and maintained, and that would impede the flow.

Operating Principle

There are two basic configurations: the curved tube flow meter and the straight tube flow meter.

The motion of the fluid relative to the axis of rotation determines what is happening.

During no-flow the outward arm and inward arm remain parallel to each other. The fluid furthest away from the axis of rotation is moving at greater velocity than the "inside track" fluid, but this doesn't require a force on radial direction.

When there is mass flow then on the outward arm fluid is moving away from the axis of rotation, and bringing it up to speed takes some pushing; the arm must exert a force on the fluid, and that makes the arm bend backwards somewhat. The inward arm on the other hand must exert a force on the fluid to decrease its velocity again, hence that arm will bend forwards.

Summarizing: During mass flow the outward arm is lagging behind the overall rotation and the inward arm is somewhat ahead.

The fluid is led through two parallel tubes. An actuator induces a vibration of the tubes. The two parallel tubes are counter-vibrating, to make the measuring device less sensitive to outside vibrations. The actual frequency of the vibration depends on the size of the mass flow meter, and ranges from 80 to 1000 vibrations per second. The amplitude of the vibration is too small to be seen, but it can be felt by touch.When no fluid is flowing, the vibration of the two tubes is symmetrical.

The vibrating mass of the setup doesn't have a constant velocity of course, but we still have that further away from the rotation axis the velocity is larger. The further away from the rotation axis, the larger the energy of the vibration.

The outward arm must exert a force on the fluid to increase vibrational energy, making it lag behind the overall vibration. The inward arm must exert a force on the fluid to decrease the fluid's vibrational energy, hence that arm gets pushed ahead of the overall vibration.

The outward arm and the inward arm vibrate with the same frequency as the overall vibration, but when there is mass flow the two vibrations shift to an out of sync pattern, the outward arm is behind, the inward arm is ahead. The two vibrations are shifted on phase with respect to each other, and the degree of phase-shift is a measure for the amount of mass that is flowing through the tubes.

References

- "A brief history of river level monitoring in Scotland". Scottish Environment Protection Agency. Retrieved 29 March 2011

- What-is-venturimeter-how-it-works: mech4study.com, Retrieved 26 March, 2019

- Miller, Richard W (1996). Flow Measurement Engineering Handbook. New York: McGraw-Hill. ISBN 978-0-07-042366-4

- Thermal-mass-flow-working-principle-theory-and-design, technical-learning: sea.omega.com, Retrieved 31 April 2019

- Explain-about-flow-nozzle-5367: instrumentationforum.com, Retrieved 09 June 2019

- Dp-flow-meter-working-application-advantages-limitation: mecholic.com, Retrieved 11 March 2019

- Applications-turbine-flowmeter: flowmetrics.com, Retrieved 08 June 2019

3
Hydrograph

The graph which shows the rate of flow of water versus time in a river channel or conduit flow is known as a hydrograph. Storm hydrograph, flood hydrograph and unit hydrograph are some of its types. The topics elaborated in this chapter will help in gaining a better perspective about these types of hydrograph.

Hydrographs are charts that display the change of a hydrologic variable over time. Here are several examples from the US Geological Survey's gaging station on the Tioga River near Mansfield, Pennsylvania. Although these examples are from a stream, hydrographs can also be made for lakes, water wells, springs and other bodies of water.

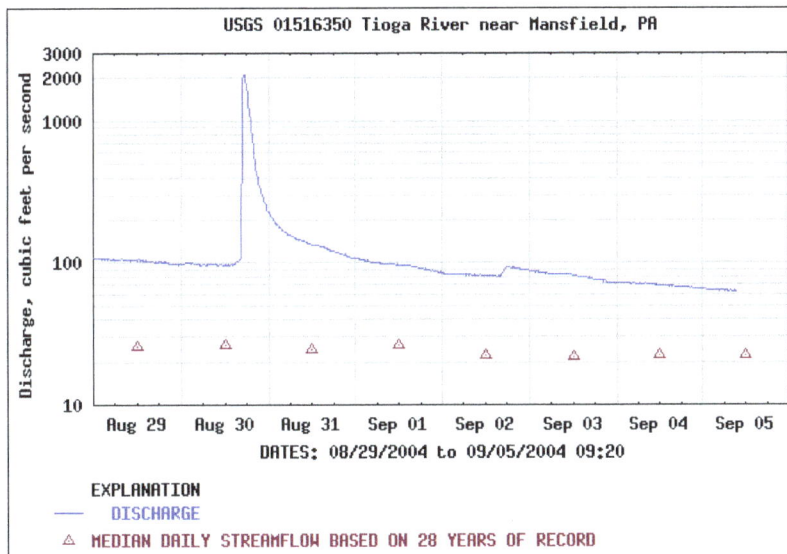

Stream discharge hydrograph.

Stream Discharge Hydrograph

This is one of the most frequently created hydrographs. It shows the change in discharge of a stream over time. The blue line on the hydrograph above shows how the discharge of the Tioga River changed between August 29 and September 5, 2004. A rainfall event in the late afternoon of August 30th produced about 1/4 inch of rain in the area of the gage. However, over one inch of rain fell in less than 15 minutes just a couple miles from the gaging station. Runoff from this precipitation caused the Tioga's discharge to rapidly increase from about 100 cubic feet per second to over 2000 cubic feet per second.

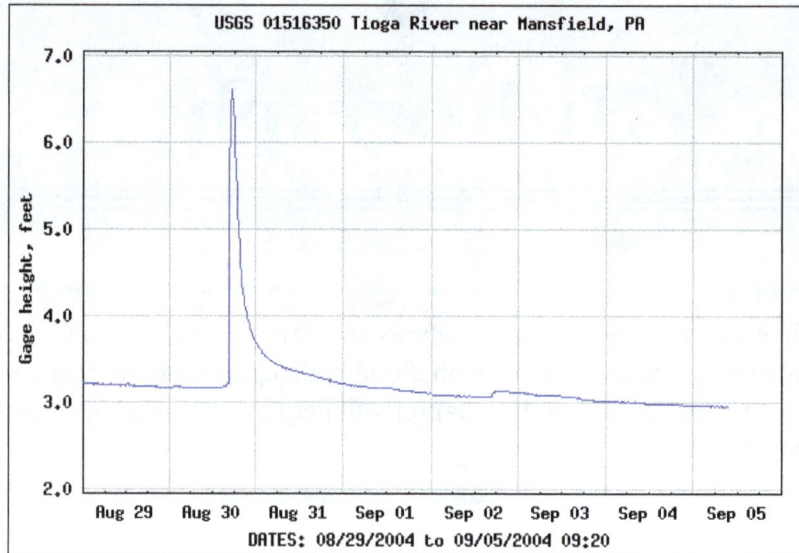

Stream stage hydrograph.

Stream Stage Hydrograph

A stream stage hydrograph shows how the height of the water above a reference datum has changed over time. Because the discharge of a stream is related to its stage, stage hydrographs and discharge hydrographs have very similar shapes.

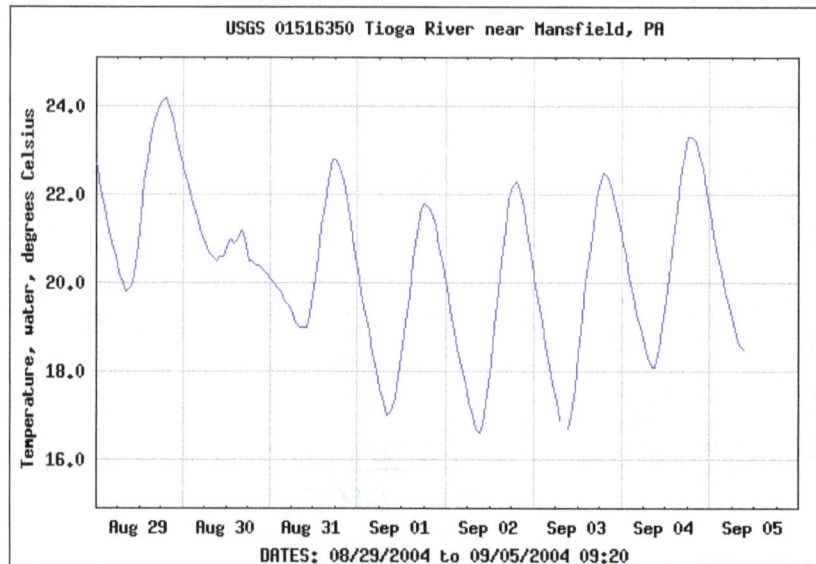

Water temperature hydrograph.

Water Temperature Hydrograph

A water temperature hydrograph shows how the temperature of the stream's water has changed over time. This water temperature hydrograph shows a daily cycle of temperature from solar heating.

When the morning sun begins heating the land, stream, and atmosphere, the water temperature begins to increase. This temperature increase continues through the day and reaches a maximum near sunset. Temperatures drop through the night, and the cycle begins again the next morning. In this hydrograph, notice how the daily temperature cycle was interrupted on August 30th. The cold precipitation/runoff lowered the water temperature and eliminated the daily temperature rise.

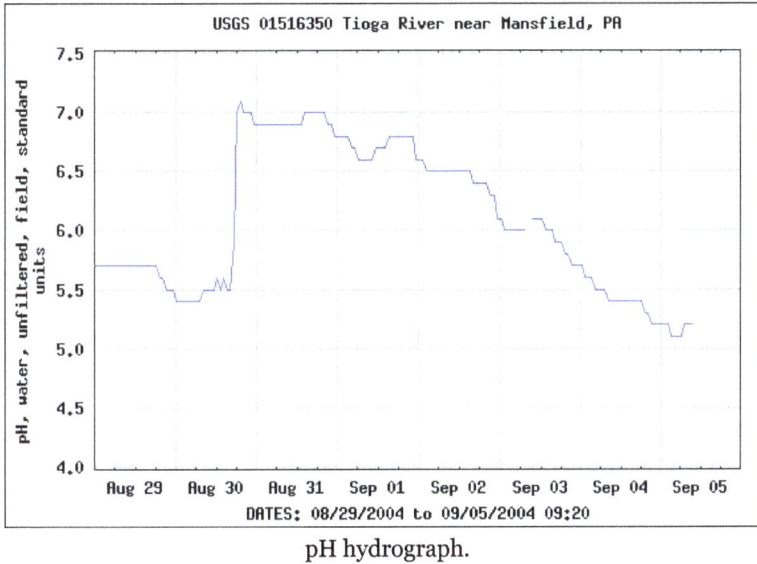

pH hydrograph.

pH Hydrograph

A pH hydrograph shows how the pH of the stream has changed over time. The pH of the Tioga River at this location is generally below 7.0. This low pH is caused by acid mine drainage entering the river at several locations upstream. The sharp increase in pH on August 30 was caused by the large amount of precipitation/runoff (probably with pH of about 7.0) entering the stream. The increased pH level slowly fell over the next few days as runoff and bank storage produced by the rainstorm slowly left the drainage area.

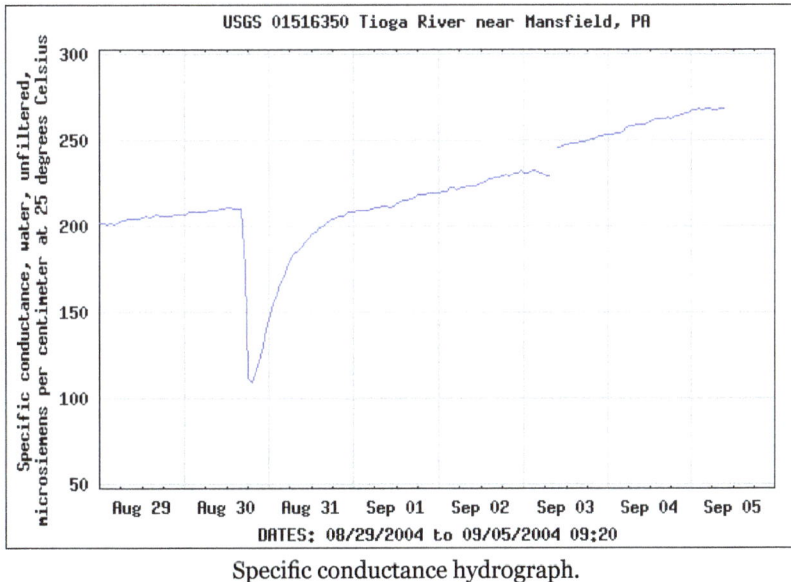

Specific conductance hydrograph.

Specific Conductance Hydrograph

Specific conductance is a measure of the ability of water to carry an electric current. This ability is proportional to the quantity of ions dissolved in the water. If you study this specific conductance hydrograph and compare it to the discharge hydrograph, you should conclude that the concentration of dissolved ions is inversely proportional to the discharge of the stream. Precipitation and runoff entering the river on August 30th diluted the concentration of dissolved ions, resulting in a sharp decrease in specific conductance. The specific conductance slowly rose through the rest of the week as runoff and bank storage water left the drainage area, and baseflow (with its high concentration of dissolved ions) began contributing a greater percentage of the stream's discharge.

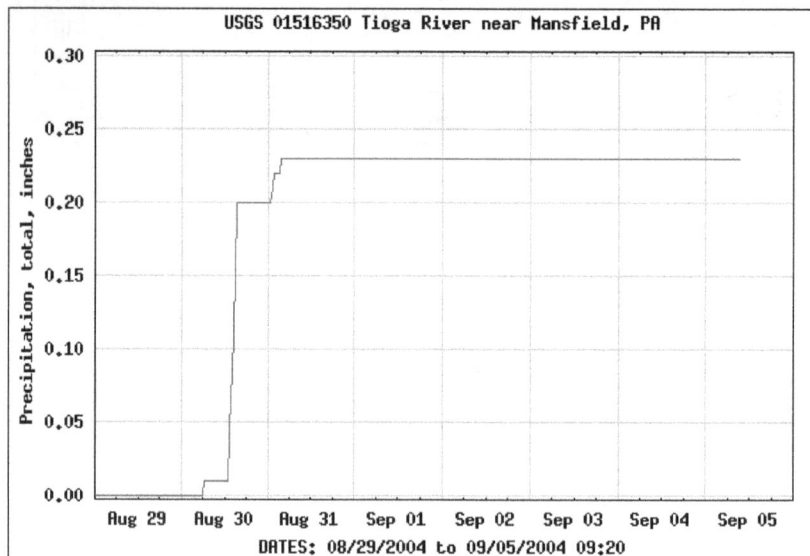

Precipitation hydrograph.

Precipitation Hydrograph

A record of precipitation over time can also be plotted as a hydrograph. The precipitation hydrograph shown is a cumulative record of precipitation at the Tioga River gaging station between August 29th and September 5th. Rainfall events are clearly shown by sharp increases of the blue line.

Components of Hydrograph

The essential components of hydrograph are illustrated in fig. and described below:

Rising Limb

The rising limb of hydrograph represents the increase in discharge due to gradual building up of storage in channels over the catchment surface. Its shape depends on the distribution of rainfall intensity, the duration, the antecedent moisture condition of the soil and the translation of runoff from the drainage basin represented by time-area diagram. At the beginning, there is only base flow (i.e., the ground water contribution to the stream) gradually depleting in an exponential form.

After the storm commences, the initial loss like interception and infiltration are met and then surface flow begins. However, they occur simultaneously because runoff starts as soon as the top soil gets saturated with water. The hydrograph gradually rises and reaches the first point of inflection where, the slope of the rising limb decreases. The first point of inflection is the point at which discharge consists of flow from all channels and after this peak is possible. the basin and storm characteristics control the shape of the rising limb of the hydrograph.

Crest Segment

The crest segment is one of the most important parts of a hydrograph as it contains the peak of the hydrograph, which represents the highest concentration of runoff from the watershed. Peak flow occurs after time to peak, which is measured from the point of rise or initiation time of rainfall to the peak flow. The time of peak with respect to initiation time of rainfall varies depending on the distribution pattern of rainfall over the watershed. Generally, the peak occurs after the rain has stopped, but sometimes, when the rainfall is of an advanced type with diminishing rainfall intensity, the peak flow occurs before the end of rainfall. Multiple peaks occur due to the development of multiple storms which are close to each other figure). After peak flow, the hydrograph further declines and reaches the second point of inflection, at which slope of the hydrograph further changes, representing the condition of maximum storage or the point at which discharge consists of no direct rain water. This is the last point of crest segment. The second point of inflection is at a distance equal to the time of concentration of the watershed from the end of excess rainfall.

Recession Limb

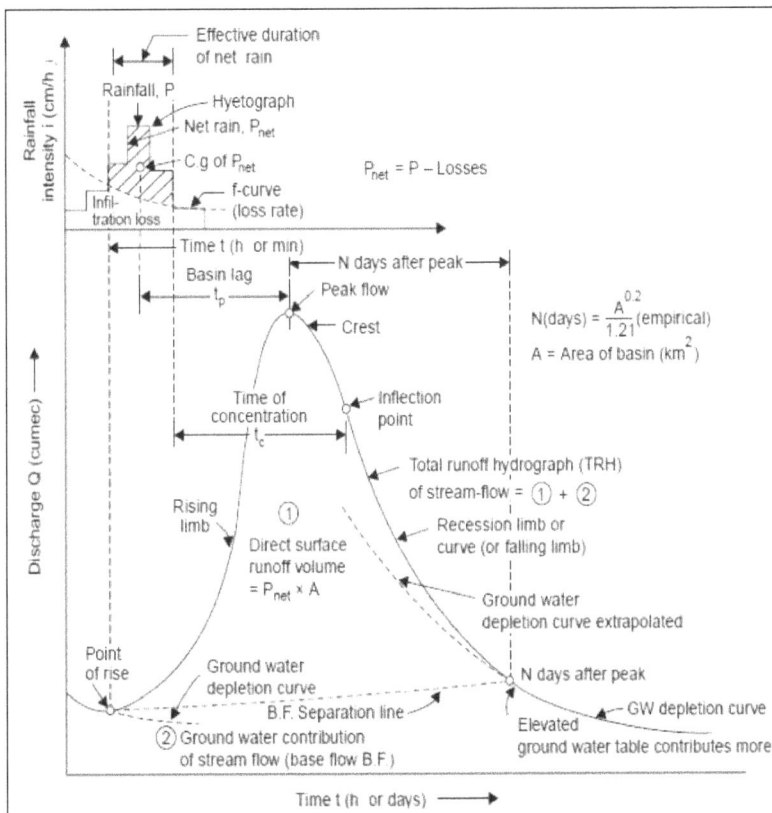

Components of flood hydrograph.

The recession limb, which extends from the second point of inflection at the end of crest segment to the commencement of the natural groundwater flow, represents the withdrawal of water from the storage in the basin during the earlier phases of hydrograph. At this time groundwater and the water held in the vadose zone contributes more into the stream flow than at the beginning of the storm. But there after the groundwater table declines and the hydrograph again goes on depleting in the exponential form called ground water depletion curve. The recession limb is independent of storm characteristics and depends entirely on the basin characteristics.

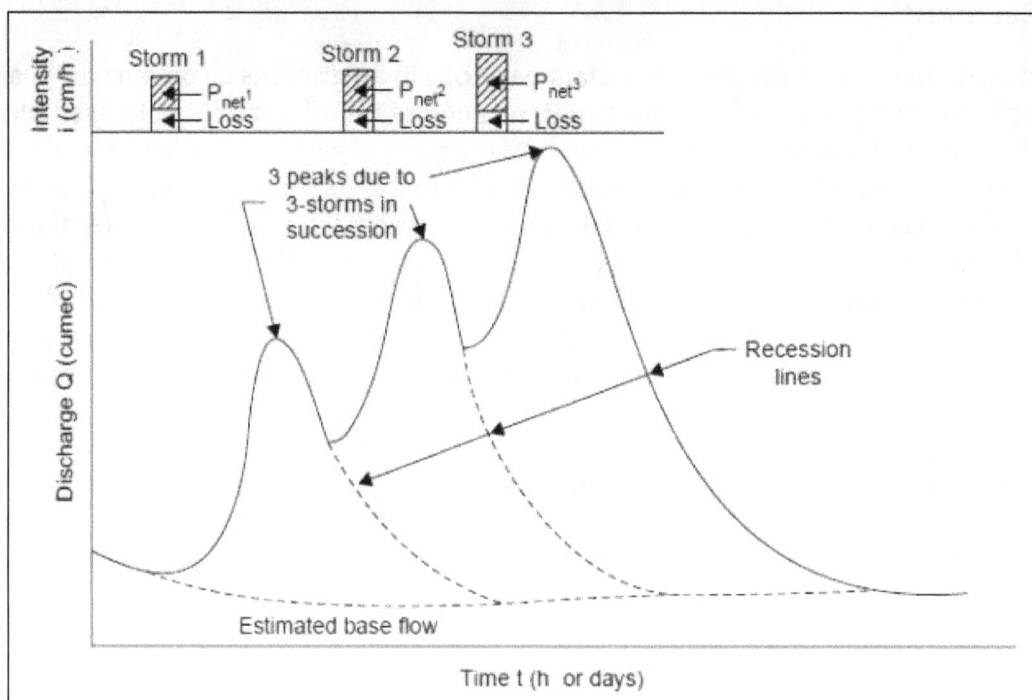

Hydrograph with multiple peaks.

Baseflow Seperation

Methods of Base Flow Separation

The surface-flow hydrograph is obtained from the total storm hydrograph by separating the quick-response flow from the slow response runoff. It is usual to consider the interflow as a part of the surface flow in view of its quick response. Thus only the base flow is to be deducted from the total storm hydrograph to obtain the surface flow hydrograph. There are three methods of base-flow separation that are in common use.

Method: In this method the separation of the base flow is achieved by joining with a straight line the beginning of the surface runoff to a point on the recession limb representing the end of the direct runoff.

In figure, point A represents the beginning of the direct runoff off and it is usually easy to identify in view of the sharp change in the runoff rate at that point. Point B, marking the end of the direct runoff is rather difficult to locate exactly. An empirical equation for the time interval N (days) from the peak to the point B is

$$N = 0.83\, A^{0.2}$$

Where A is drainage area in km2and N is in days. Points A and B are joined by a straight line to demarcate to the base flow and surface runoff. This method of base-flow separation is the simplest of all the three methods.

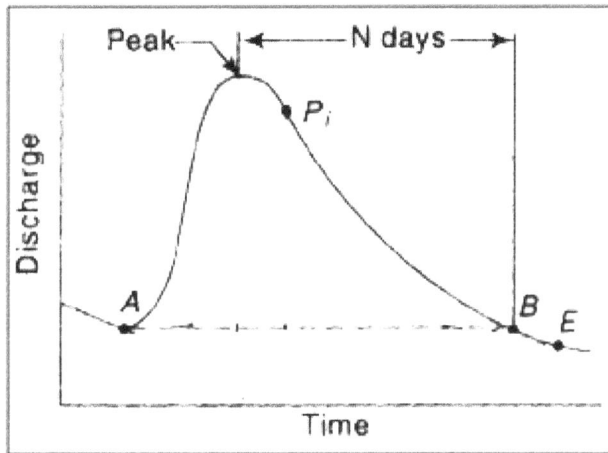

Method 1 for base flow separation.

Method: In this method the base flow curve existing prior to the commencement of the surface runoff is extended till it intersects the ordinate drawn at the peak (point C in figure). This point is joined to point B by a straight line. Segment AC and CB demarcate the base flow and surface run-off. This is probably the most widely used base-flow separation procedure.

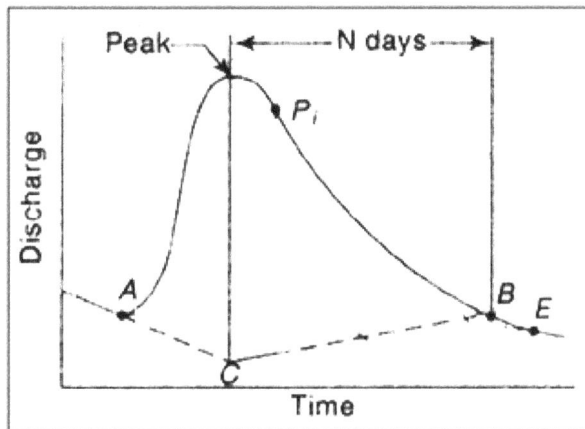

Method 2 for base flow separation.

Method: In this method the base flow recession curve after the depletion of the flood water is extended backwards till it intersects the ordinate at the point of inflection (line EF in figure). Points A and F are joined by an arbitrary smooth curve. This method of base-flow separation is realistic in situations where the groundwater contributions are significant and reach the stream quickly.

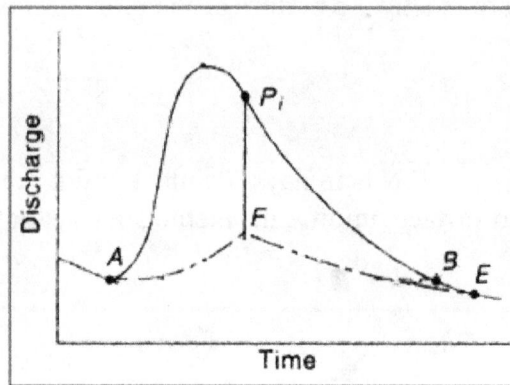

Method 3 for base flow separation.

The surface runoff hydrograph obtained after the base-flow separation is also known as direct runoff hydrograph (DRH).

Example: The following are the ordinates of the hydrograph of flow from a catchment area of 770 km² due to a 6-h rainfall. Derive the ordinates of DRH. Make suitable assumptions regarding the base flow.

Time from beginning of storm	(h)	0	6	12	18	24	30	36
Discharge	(m³/s)	42	65	215	360	400	350	270
Time from beginning of storm	(h)	42	48	54	60	66	72	
Discharge	(m³/s)	205	145	100	70	50	42	

Answer: Given, Catchment area (A) = 770 km²

Using equation $N = 0.83\,A^{0.2}$,

$$N = 0.83\,A^{0.2}$$

$$N = 0.83\left(770\right)^{0.2}$$

$$N = 3.13 \text{ day}$$

$$N = 75.26\ h \quad \textit{from peak}$$

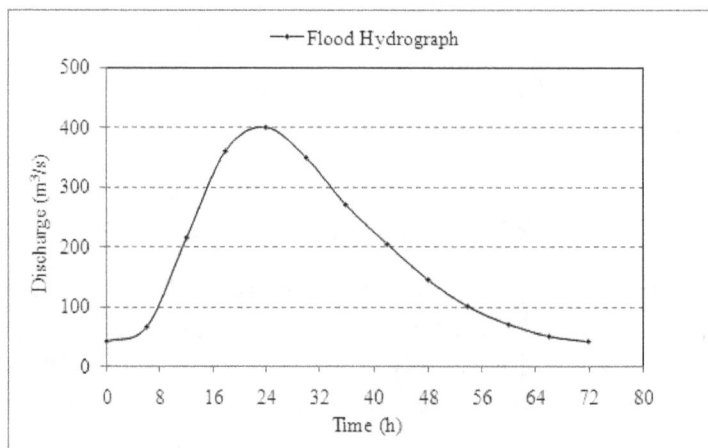

From given data, with our convenience, base flow = 42 m³/s at 72 h.

Therefore,

DRH = Flood Hydrograph – Base flow

Time from beginning of storm	Discharge	Base flow	DRH
h	m³/s	m³/s	m³/s
0	40	42	-2
6	65	42	23
12	215	42	173
18	360	42	318
24	400	42	358
30	350	42	308
36	270	42	228
42	205	42	163
48	145	42	103
54	100	42	58
60	70	42	28
66	50	42	8
72	42	42	0

Example: The daily stream flow data at a site having a drainage area of 6500 km² are given in the following table. Separate the base flow using the above three methods.

Time (days)	Discharge (m3/s)
1	1600
2	1550
3	5000
4	11300
5	8600
6	6500
7	5000
8	3800
9	2800
10	2200
11	1850
12	1600
13	1330
14	1300
15	1280

Answer: 1. Plot the total runoff hydrograph method: join point A, the beginning of direct runoff, to point B, the end of direct runoff. Both points are selected by judgment.

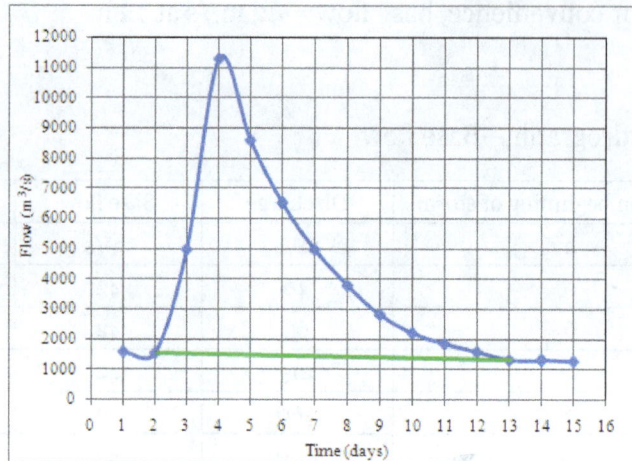

2. Method: Extend the recession curve before the storm up to point C below the peak. Join point C to D, computed using equation:

$$N = 0.83 \ A^{0.2} = 0.83(6500)^{0.2} = 4.6 \ days \ (Approx. 5 \ days)$$

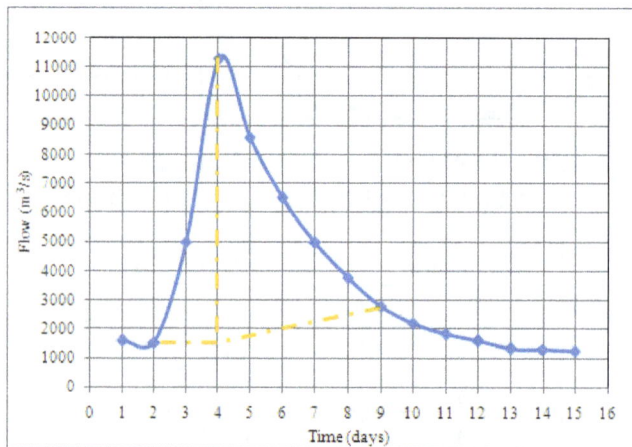

3. Method: Extend the recession curve backward to point E. Join point E to A:

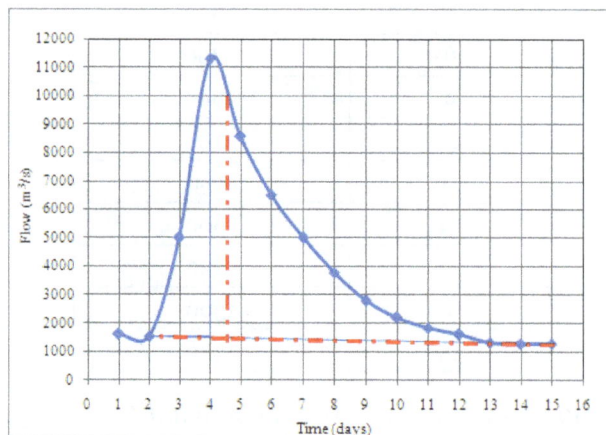

4. The ordinates DRH by three methods are given in table.

Table: Ordinates of DRH by different methods.

Time	Total runoff	Base flow			Direct runoff		
		Method 1	Method 2	Method 3	Method 1	Method 2	Method 3
(days)	(m³/s)	(m³/s)	(m³/s)	(m³/s)	(m³/s)	(m³/s)	(m³/s)
1	1600	1600	1600	1600	0	0	0
2	1550	1550	1550	1550	0	0	0
3	5000	1520	1480	1500	3480	3520	3500
4	11300	1500	1400	1450	9800	9900	9850
5	8600	1450	1700	1400	7150	6900	7200
6	6500	1450	1950	1400	5050	4550	5100
7	5000	1450	2300	1400	3550	2700	3600
8	3800	1400	2550	1400	2400	1250	2400
9	2800	1380	2800	1380	1420	0	1420
10	2200	1380	2200	1380	820	0	820
11	1850	1380	1850	1380	470	0	470
12	1600	1350	1600	1350	250	0	250
13	1330	1330	1330	1330	0	0	0
14	1300	1300	1300	1300	0	0	0
15	1280	1280	1280	1280	0	0	0

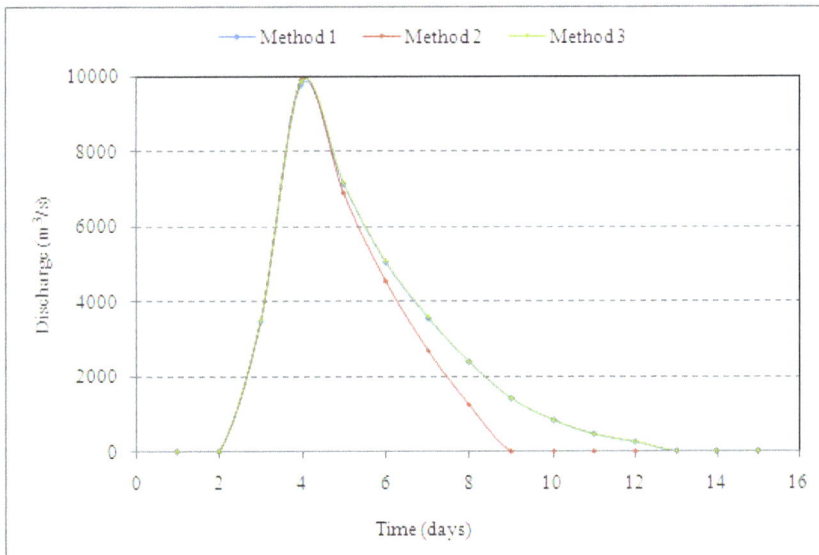

Effective Rainfall Hyetograph

Effective rainfall (also known as Excess rainfall) (ER) is that part of the rainfall that becomes direct runoff at the outlet of the watershed. It is thus the total rainfall in a given duration from which abstractions such as infiltration and initial losses are subtracted.

For purposes of correlating DRH with the rainfall which produced the flow, the hyetograph of the rainfall is also pruned by deducting the losses. Figure shows the hyetograph of a storm. The initial loss and infiltration losses are subtracted from it. The resulting hyetograph is known as effective rainfall hyetograph (ERH). It is also known as excess rainfall hyetograph.

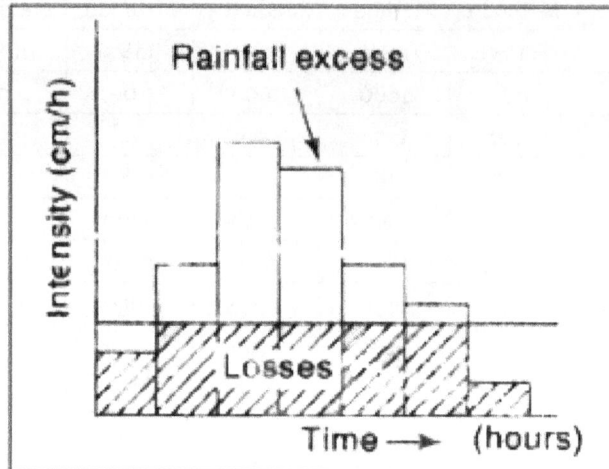

Effective rainfall hyetograph.

Both DRH and ERH represent the same total quantity but in different units. Since ERH is usually in cm/h plotted agains1 time, the area of ERH multiplied by the catchment area gives the total volume of direct runoff which is the same as the area of DRH. Theinitial loss and infiltration losses are estimated based on the available data of the catchment.

Example: A 4-hour storm occurs over an 80 km2 watershed. The details of the catchment are as follows.

Sub Area	Φ index	Hourly rain (mm)			
km²	mm/h	1ˢᵗ hour	2ⁿᵈ hour	3ʳᵈ hour	4ᵗʰ hour
15	10	16	48	22	10
25	15	16	42	20	8
35	21	12	40	18	6
5	16	15	42	18	8

Calculate the runoff from catchment and the hourly distribution of the effective rainfall whole catchment.

Answer:

$$= \frac{(16-10)\times(16-15)\times 25 + (12-21)\times 35(15-16)\times 5}{15+25+35+5}$$

$$= \frac{6\times 15 + 1\times 25 + 0\times 35 + 0\times 5}{80}$$

$$= 1.4375 \; mm$$

2nd hour,

$$= \frac{(48-10)\times15+(42-15)\times25+(40-21)\times35+(42-16)\times5}{15+25+35+5}$$

$$= \frac{38\times15+27\times25+19\times35+24\times5}{80}$$

$$= 25.375 \ mm$$

3rd hour,

$$= \frac{(22-10)\times15+(20-15)\times25+(18-21)\times35+(18-16)\times5}{15+25+35+5}$$

$$= \frac{12\times15+5\times25+0\times35+2\times5}{80}$$

$$= 3.9375 \ mm$$

4th hour,

$$= \frac{(10-10)\times15+(8-15)\times25+(6-21)\times35+(8-6)\times5}{15+25+35+5}$$

$$= \frac{0\times15+0\times25+0\times35+0\times5}{80}$$

$$= 0 \ mm$$

Total runoff = 1.4375 + 25.375 + 3.9375 = 30.75 mm

$$\text{Total runoff} = \left(\frac{30.75}{1000}\right)\times80\times10^8$$

Totalrunoff = 2.46Mm³

Hourly distribution of the effective rainfall for the whole catchment:

	Effective rainfall (mm)
1st hour	1.4375
2nd hour	25.375
3rd hour	0
4th hour	3.9375

Example:

A storm in a certain catchment had three successive 6-h intervals of rainfall magnitude of 3.0 cm, 5.0 cm and 4.0 cm, respectively.

The flood hydrograph at the outlet of the catchment resulting from this storm is as follows:

Time	(h)	0	6	12	18	24	30	36	42
Flood hydrograph ordinates	(m³/s)	30	480	2060	4450	6010	6010	5080	3996
Time	(h)	48	54	60	66	72	78		
Flood hydrograph ordinates	(m³/s)	2866	1866	1060	500	170	30		

If the area of the catchment is 8791.2 km², estimate the index of the storm. Assume the base flow as 30 m³/s.

Answer: Flood hydrograph ordinates = DRH ordinates + Base flow ordinates

$$\text{Direct runoff(cm)} = 0.36\frac{\left(\sum_{i=1}^{N} DRO_i\right)\Delta t}{A}$$

Where,

is direct runoff ordinates (m³/s), is time interval between successive ordinates (h), A is catchment area (km²)

Time	Flood hydrograph Ordinates	Base flow	DRO
h	m³/s	m³/s	m³/s
0	30	30	0
6	480	30	450
12	2060	30	2030
18	4450	30	4420
24	6010	30	5980
30	6010	30	5980
36	5080	30	5050
42	3996	30	3966
48	2866	30	2836
54	1866	30	1836
60	1060	30	1030
66	500	30	470
72	170	30	140
78	30	30	0
		= 34188	

Therefore,

$$\text{Direct runoff (cm)} = 0.36 \times \frac{34188 \times 6}{8791.2}$$

Directrunoff (cm) = 8.4

Therefore

$$(3-6\varphi)+(5-6\varphi)+(4-6\varphi)=8.4$$

$\varphi = 0.2\text{cm/h}$

Rainfall (cm)	3	5	4
Time interval (h)	6	6	6
Rainfall intensity (cm/h)	0.5	0.833	0.667
index (cm/h)	0.2	0.2	0.2
Excess rainfall intensity	0.3	0.633	0.467

Elemental Hydrograph

If a small, impervious area is subjected to a constant rate rainfall, the resulting runoff hydrograph will appear much as above, and is known as elemental hydrograph. In the beginning, there will be surface detention (rainfall-runoff) so as to start the sheet flow over the surface. At point B, known as point of equilibrium, outflow rate equals inflow rate. When rainfall ends (at C), recession starts, i.e., outflow rate and detention volume increases.

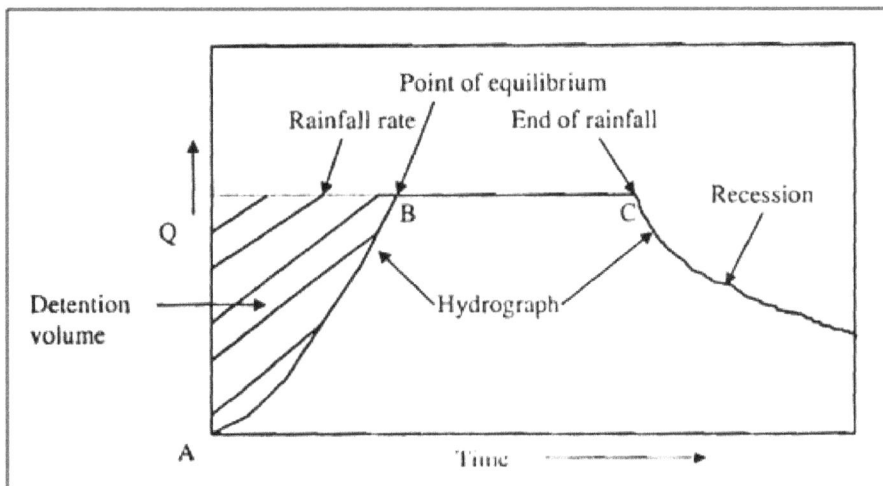

Elemental hydrograph.

Storm Hydrograph

Storm hydrographs are graphs that show how a drainage basin responds to a period of rainfall. They are useful in planning for flood situations and times of drought as they show the discharge (amount of water reaching channel via surface run-off, throughflow, and base flow) that originated as precipitation.

Reading the Hydrograph

A great deal of information can be gleaned from a hydrograph and the interpretation of them is often tested in exam questions. The diagram below shows the main points:

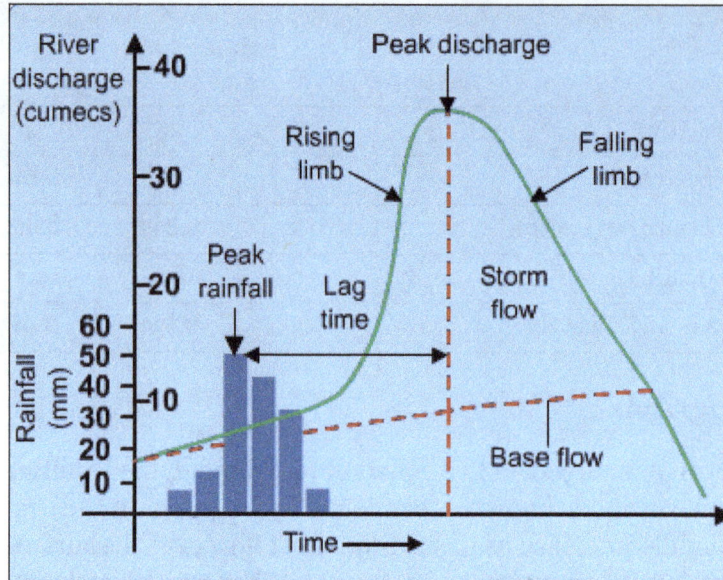

Influences on the Hydrographs and Drainage Basin

Drainage basins all have a variety of characteristics in terms of vegetation, geology, soil type and so on, all of which interact to influence how quickly or slowly river discharge increases after a storm.

The major influences on hydrographs and drainage basins are given below:

Size of Basin, Shape and Relief

Size - the smaller the basin the less time it takes for water to drain to the river, resulting in a shorter lag time. Shape - the shape of basin that lends itself to most rapid drainage is circular. In a long, narrow basin water takes longer to reach the river. Relief - the steeper the basin the more quickly it drains.

Forms of Precipitation

Heavy Storms - in such a situation, rainfall is often far in excess of the infiltration capacity of the soil leading to much overland flow, and rapid rises in river levels. Lengthy rainfall - leads to the ground being saturated and overland flow. Snowfall - until snow melts, potential discharge for a river is held in storage. Rapid melting can lead to flooding.

Temperature

High rates of evapotranspiration reduce amounts of discharge, and low temperatures can store water in the form of ice and snow.

Land Use

Vegetation - Important in reducing discharge as it intercepts precipitation and adds to rates of evapotranspiration. Roots of plants take up water reducing throughflow. Interception is less in winter in the UK due the shedding of leaves from deciduous trees. Flooding is more likely in deforested areas.

Geology

Rock type varies within drainage basins and can be permeable (allowing water through) or impermeable (not allowing water through). Permeable rocks can be porous such as chalk that store water within them or pervious, such as limestone where water flows along bedding plains. Impermeable rocks encourage grater amounts of surface run-off and a more rapid increase in discharge than permeable rocks.

Soil

A control on the rate of infiltration, amount of soil moisture storage and rate of throughflow. Larger pore spaces as found in sand, allow for greater water storage and limit the risk of flooding.

Drainage Density

The higher the density the greater the risk of flooding.

Tides and Storms

High spring tides (illustrated by the Severn Bore) prevent water from entering the sea and increase the risk of flooding.

Urbanisation

A major impact because of its alteration on the hydrological process.

Flood Hydrograph

A flood hydrograph shows the amount of rainfall in an area and the discharge of a river. The discharge of a river is the volume of water passing a point each second. It is expressed in cumecs (cubic metres per second). River discharge is displayed as a line graph. Precipitation is shown as a bar graph and is usually displayed in millimetres.

The starting and finishing level show the base flow of a river. The base flow is the water that reaches the channel through slow throughflow and permeable rock below the water table. As stormwater enters the drainage basin the discharge rates increase. This is shown in the rising limb. The highest flow in the channel is known as the peak discharge. The fall in discharge back to base level is shown in the preceding limb. The lag time is the delay between the maximum rainfall amount and the peak discharge. The shape of a hydrograph varies in each river basin and each individual storm event.

Flood hydrographs can be used to predict flooding by showing how different levels of precipitation affect a river during a storm. Hydrographs can be different shapes. The characteristics of the river and how likely it is to flood affects its shape.

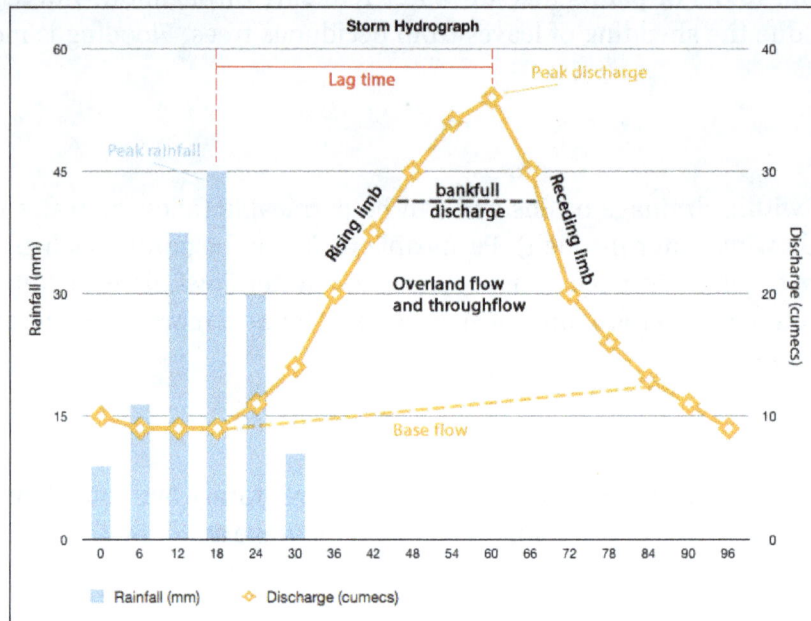

A storm hydrograph.

A gentle hydrograph shows the river is at low risk of flooding. These types of hydrograph have a gentle rising limb and a long lag time which means it takes longer for the peak rainfall to reach the river channel, so the river discharge is increasing slowly.

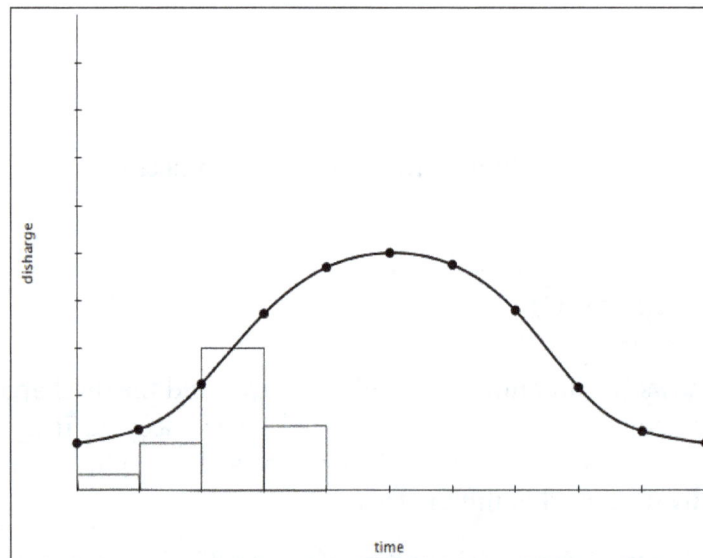

A gentle hydrograph.

Flashy hydrographs have a steep rising limb and a small lag time. This indicates the river discharge increases rapidly over a short period of time, indicating rainwater reaches the river very quickly. This means the river is more likely to flood.

Rural areas with predominantly permeable rock increase infiltration and decreases surface runoff. This increases lag time. The peak discharge is also lower as it takes water longer to reach the river channel.

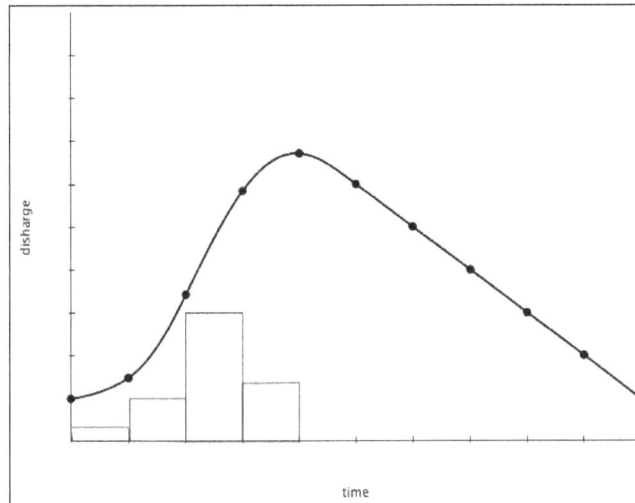

A flashy hydrograph.

Physical Factors affecting Storm Hydrographs

There is a range of physical factors that affect the shape of a storm hydrograph. These include:

- Large drainage basins catch more precipitation so have a higher peak discharge compared to smaller basins. Smaller basins generally have shorter lag times because precipitation does not have as far to travel. The shape of the drainage basin also affects runoff and discharge. Drainage basins that are more circular in shape lead to shorter lag times and a higher peak discharge than those that are long and thin because water has a shorter distance to travel to reach a river.

- Drainage basins with steep sides tend to have shorter lag times than shallower basins. This is because water flows more quickly on the steep slopes down to the river.

- Basins that have many streams (high drainage density) drain more quickly so have a shorter lag time.

- If the drainage basin is already saturated then surface runoff increases due to the reduction in infiltration. Rainwater enters the river quicker, reducing lag times, as surface runoff is faster than baseflow or through flow.

- If the rock type within the river basin is impermeable surface runoff will be higher, throughflow and infiltration will also be reduced meaning a reduction in lag time and an increase in peak discharge.

- If a drainage basin has a significant amount of vegetation this will have a significant effect on a storm hydrograph. Vegetation intercepts precipitation and slows the movement of water into river channels. This increases lag time. Water is also lost due to evaporation and transpiration from the vegetation. This reduces the peak discharge of a river.

- The amount of precipitation can have an effect on the storm hydrograph. Heavy storms result in more water entering the drainage basin which results in a higher discharge. The type of precipitation can also have an impact. The lag time is likely to be greater if the precipitation is snow rather than rain. This is because snow takes time to melt before the water enters the river channel. When there is rapid melting of snow the peak discharge could be high.

Human Factors affecting Storm Hydrographs

There is a range of human factors that affect the shape of a storm hydrograph. These include:

1. Drainage systems that have been created by humans lead to a short lag time and high peak discharge as water cannot evaporate or infiltrate into the soil.

2. Areas that have been urbanised result in an increase in the use of impermeable building materials. This means infiltration levels decrease and surface runoff increases. This leads to a short lag time and an increase in peak discharge.

Unit Hydrograph

The unit time or unit hydrograph duration is the duration for occurrence of precipitation excess. The optimum unit time is less than 20 percent of the time interval between the beginning of runoff from a short duration, high-intensity storm and the peak discharge of the corresponding runoff.

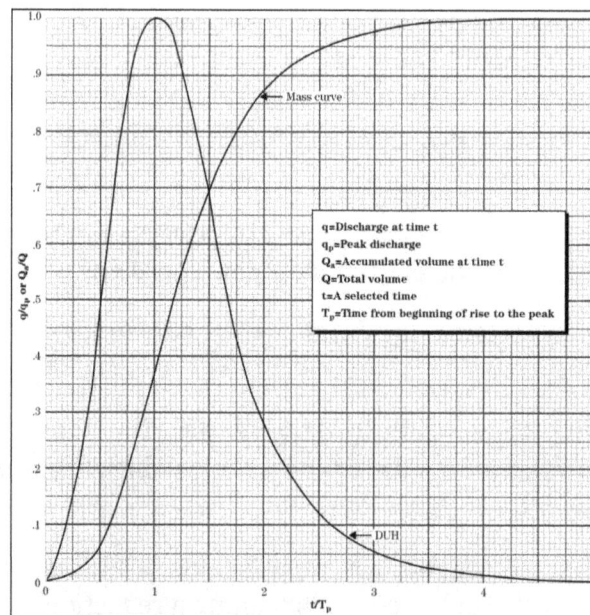

q=Discharge at time t
q_p=Peak discharge
Q_a=Accumulated volume at time t
Q=Total volume
t=A selected time
T_p=Time from beginning of rise to the peak

Dimensionless unit hydrograph and mass curve.

The storm duration is the actual duration of the precipitation excess. The duration varies with actual storms. The dimensionless unit hydrograph used by the Natural Resources Conservation Service (NRCS) was developed by Victor Mockus. It was derived from many natural unit hydrographs from watersheds varying widely in size and geographical locations. This dimensionless curvilinear hydrograph, has its ordinate values expressed in a dimensionless ratio q/q_p or Q_a/Q and its abscissa values as t/T_p. This unit hydrograph has a point of inflection approximately 1.7 times the time to peak (T_p).

Elements of a Unit Hydrograph

The dimensionless curvilinear unit hydrograph has 37.5 percent of the total volume in the rising side, which is represented by one unit of time and one unit of discharge. This dimensionless unit hydrograph also can be represented by an equivalent triangular hydrograph having the same units of time and discharge, thus having the same percent of volume in the rising side of the triangle. This allows the base of the triangle to be solved in relation to the time to peak using the geometry of triangles. Solving for the base length of the triangle, if one unit of time T_p equals 0.375 per cent of volume:

$$T_b = \frac{1.00}{.375} = 2.67 \quad \text{units of time}$$

$$T_r = T_h - T_p = 1.67 \quad \text{units of time or } 1.67\ T_p$$

where,

T_b = time from beginning to end of the triangular hydrograph.

T_r = time from the peak to the end of the triangular hydrograph.

T_p = time from the beginning of the triangular hydrograph to its peak.

These relationships are useful in developing the peak rate equation for use with the dimensionless unit hydrograph.

Peak Rate Equation

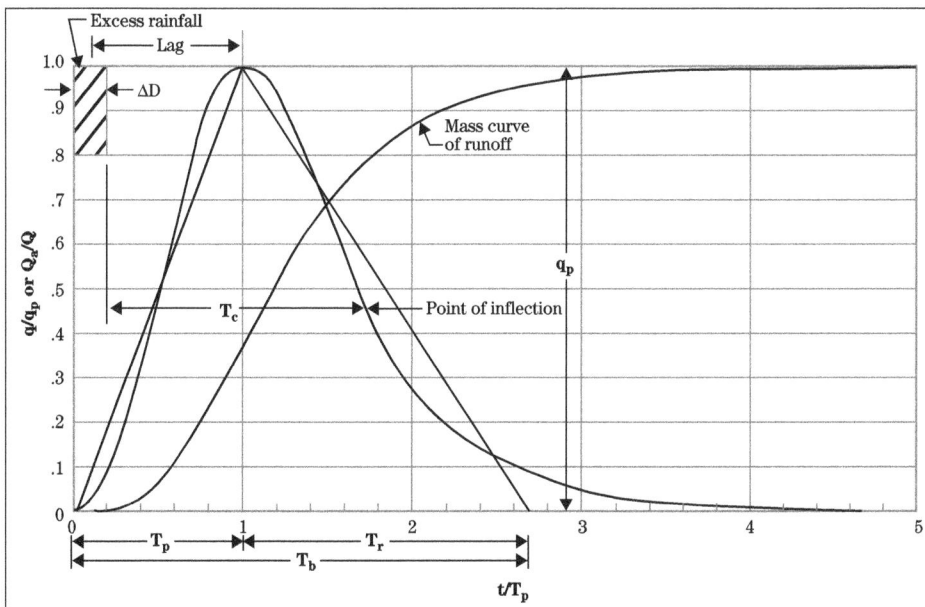

Dimensionless curvilinear unit hydrograph and equivalent triangular hydrograph.

From figure the total volume under the triangular unit hydrograph is:

$$Q = \frac{q_p T_p}{2} + \frac{q_p T_r}{2} = \frac{q_p}{2}\left(T_p + T_r\right)$$

With Q in inches and T in hours, solve for peak rate q_p in inches per hour:

$$q_p = \frac{2Q}{T_p + T_r}$$

Let:

$$K = \frac{2}{1 + \dfrac{T_r}{T_p}}$$

Therefore:

$$q_p = \frac{KQ}{T_p}$$

In making the conversion from inches per hour to cubic feet per second and putting the equation in terms ordinarily used, including drainage area A in square miles and time T in hours, equation $q_p = \dfrac{KQ}{T_p}$ becomes the general equation:

$$q_p = \frac{645.33 \times K \times A \times Q}{T_p}$$

where,

q_p = peak discharge in cubic feet per second (ft³/s)

645.33 = conversion factor for the rate required to discharge 1 inch from 1 square mile in 1 hour, units of

$$\frac{\left(ft^3 / s\right) \times h}{mi^2 \times in}$$

K = nondimensional factor

A = drainage area in square miles (mi²)

Q = runoff in inches (in)

T_p = time to peak in hours (h)

The relationship of the triangular unit hydrograph, Tr = 1.67, gives K = 0.75. Then substituting into equation $q_p = \dfrac{645.33 \times K \times A \times Q}{T_p}$ gives:

$$q_P = \frac{484\, AQ}{T_p}$$

The peak rate factor for the triangular dimensionless unit hydrograph is 484. The curvilinear dimensionless unit hydrograph of table has a peak rate factor of 483.4 due to rounding of the discharge ratios to either two or three decimal places. This discrepancy produces a negligible difference in the calculated peak discharges.

Any change in the dimensionless unit hydrograph reflecting a change in the percent of volume under the rising side would cause a corresponding change in the shape factor associated with the triangular hydrograph and, therefore, a change in the constant 484. This constant has been known to vary from about 600 in steep terrain to 100 or less in flat, swampy country. The national hydraulic engineer should concur in the use of a dimensionless unit hydrograph other than that shown in figure. If the dimensionless shape of the hydrograph needs to vary to perform a special job, the ratio of the percent of total volume in the rising side of the unit hydrograph to the rising side of a triangle is a useful tool in arriving at the peak rate factor.

Figure shows that:

$$T_p = \frac{\Delta D}{2} + L$$

where:

ΔD = duration of unit excess rainfall in hours

L = watershed lag in hours

The lag of a watershed is defined in NEH 630 as the time from the center of mass of excess rainfall $\left(\dfrac{\Delta D}{2}\right)$ to the time to peak (T_p) of a unit hydrograph. Combining equations $q_P = \dfrac{484\, AQ}{T_p}$ and $T_p = \dfrac{\Delta D}{2} + L$ results in equation $q_p = \dfrac{484\, AQ}{\dfrac{\Delta D}{2} + L}$:

$$q_p = \frac{484\, AQ}{\dfrac{\Delta D}{2} + L}$$

The average relationship of lag to time of concentration (T_c) is L = 0.6 T_c. T_c is expressed in hours.

Substituting in equation $q_p = \dfrac{484\,AQ}{\dfrac{\Delta D}{2} + L}$, the peak rate equation becomes:

$$q_p = \frac{484\,AQ}{\dfrac{\Delta D}{2} + 0.6\,T_c}$$

The time of concentration is defined in two ways

- The time for runoff to travel from the hydraulically most distant point in the watershed to the point in question.

- The time from the end of excess rainfall to the point of inflection on the recedimg limb of the unit hydrograph.

These two relationships are important because T_c is computed under the first definition and ΔD, the unit storm duration, is used to compute the time to peak (T_p) of the unit hydrograph. This in turn is applied to all of the points on the abscissa of the dimensionless unit hydrograph using the ratio t/T_p as shown in table.

The dimensionless unit hydrograph shown in figure has a time to peak at one unit of time and point of inflection at approximately 1.7 units of time. Using the relationships Lag = 0.6 T_c and the point of inflection = 1.7 T_p, ΔD will be 0.2 Tp. A small variation in DD is permissible; however, it should be no greater than 0.25 T_p. The standard NRCS dimensionless unit hydrograph is defined at an interval of 0.1 T_p. If ΔD is 0.2 T_p, then during computations, every second point is used to develop the incremental flood hydrograph. If ΔD is 0.25 T_p or more, then the shape of the dimensionless unit hydrograph is not being represented accurately.

Using the relationship shown on the dimensionless unit hydrograph in figure, compute the relationship of ΔD to T_c:

$$T_c + \Delta D = 1.7\,T_p$$

$$\frac{\Delta D}{2} + 0.6\,T_c = T_p$$

Combining these two equations results in a defined relationship between ΔD and T_c:

$$T_c + \Delta D = 1.7\left(\frac{\Delta D}{2} + 0.6\,T_c\right)$$

$$0.15\,\Delta D = 0.02\,T_c$$

$$\Delta D = 0.133\,T_c$$

Applications of Unit Hydrograph

The unit hydrograph can be constructed for any location on a regularly shaped watershed, once the values of q_p and T_p are defined (fig., areas A and B).

Area C in figure is an irregularly shaped watershed having two regularly shaped areas (C2 and C1) with a large difference in their time of concentration. This watershed requires the development of two unit hydrographs that may be added together, forming one irregularly shaped unit hydrograph. This irregularly shaped unit hydrograph may be used to develop a flood hydrograph in the same way as the unit hydrograph developed from the dimensionless form is used to develop the flood hydrograph. Example which develops a composite flood hydrograph for area A shown in figure. Also, each of the two unit hydrographs developed for areas C2 and C1 in figure may be used to develop flood hydrographs for the respective areas C2 and C1. The flood hydrographs from each area are then combined to form the hydrograph at the outlet of area C.

Many variables are integrated into the shape of a unit hydrograph. Since a dimensionless unit hydrograph is used and the only parameters readily available from field data are drainage area and time of concentration, consideration should be given to dividing the watershed into hydrologic units of uniformly shaped areas. These subareas, if at all possible, should have a homogeneous drainage pattern, homogeneous land use and approximately the same size. To assure that all contributing subareas are adequately represented, it is suggested that no subarea exceed 20 square miles in area and that the ratio of the largest to the smallest drainage area not exceed 10.

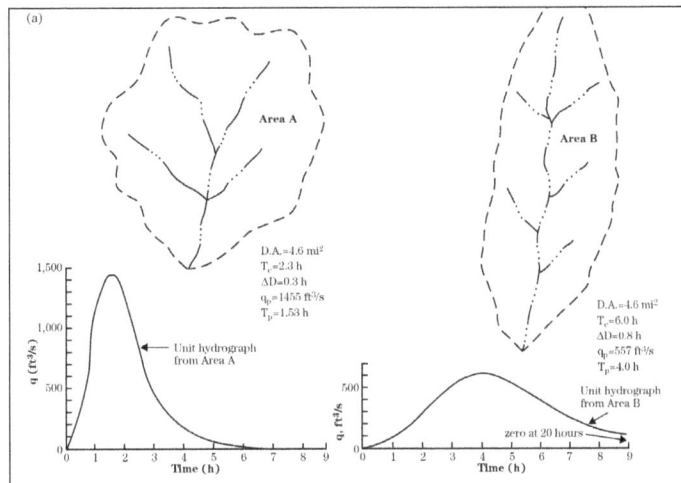

Effect of watershed shape on the peaks of unit hydrographs (Equations, definitions, and units for variables are found in appendix 16A).

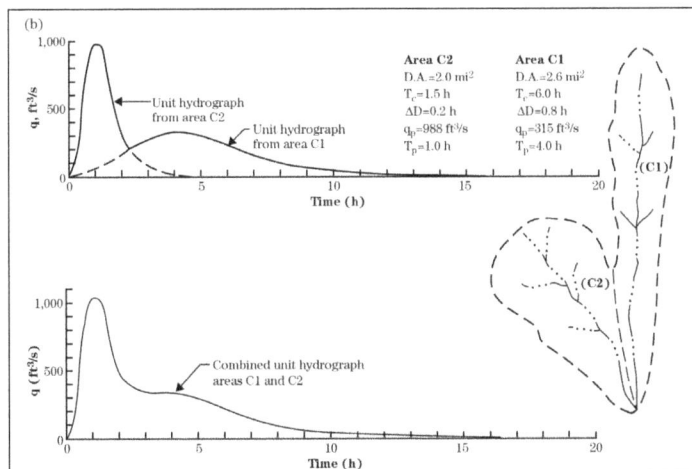

Effect of watershed shape on the peaks of unit hydrographs.

Annual Hydrograph or Regimes

River regime can describe one of two characteristics of a reach of an alluvial river:

- The variability in its discharge throughout the course of a year in response to precipitation, temperature, evapotranspiration and drainage basin characteristics.

- A series of characteristic power-law relationships between discharge and width, depth and slope.

The latter is described by the fact that the discharge through a river of an approximate rectangular cross-section must, through conservation of mass, equal:

$$Q = \bar{u}bh$$

where Q is the volumetric discharge, $\bar{u}$ is the mean flow velocity, is the channel width (breadth) and h is the channel depth.

Because of this relationship, as discharge increases, depth, width, and/or mean velocity must increase as well.

Empirically-derived relationships between depth, slope, and velocity are:

$$b \propto Q^{0.5}$$
$$h \propto Q^{0.4}$$
$$u \propto Q^{0.1}$$

Q refers to a "dominant discharge" or "channel-forming discharge", which is typically the 1–2 year flood, though there is a large amount of scatter around this mean. This is the event that causes significant erosion and deposition and determines the channel morphology.

The variability in discharge over the course of a year is commonly represented by a hydrograph with mean monthly discharge variations plotted over the annual time scale. When interpreting such records of discharge, it is important to factor in the time scale over which the average monthly values were calculated. It is particularly difficult to establish a typical annual river regime for rivers with high interannual variability in monthly discharge and/or significant changes in the catchment's characteristics (e.g. tectonic influences or the introduction of water management practices).

Classification

There are three basic types of regimes:

- Simple regime - one maximum and one minimum per year

- Mixed regime/double regime - two maximums and two minimums per year

- Complex mode - several extrema

Simple Regimes

Simple regimes can be nival, pluvial or glacial, depending on the origin of the water. Simple regime is where all rivers have one peak discharge per year

Glacial Regime

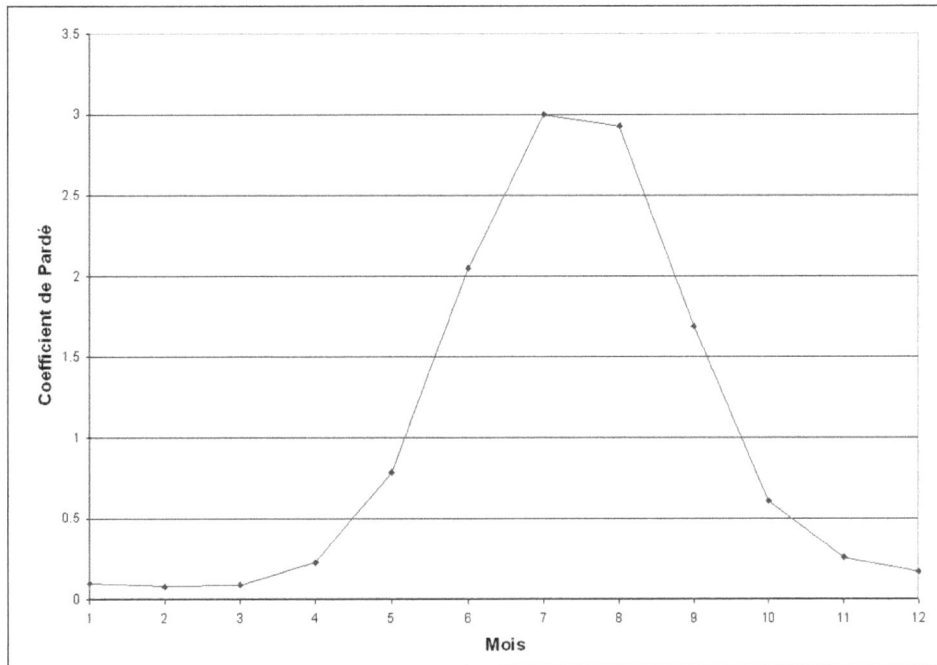

Glacial regime of the Kander (Switzerland).

The glacial regime is characterised by:

- Very high discharge in summer after the ice melt.

- Very low discharge from the end of autumn to early spring.

- Amplitude of monthly variation of discharge greater than 25.

- Very high daily variability in discharge during the year.

- High flow (several hundred l/s/km2).

It is found at high altitudes, above 2,500 metres (8,200 ft).

Example: Rhône at Brigue.

Nival

The nival regime is similar to the glacial, but attenuated and the maximum takes place earlier, in June. It can be mountain or plain nival. The characteristics of the plain nival (example: Simme at Oberwi) are:

- Short and violent flood in April–May following massive spring thawing ofwinter snows.

- Great daily variability.
- Very great variability over the course of the year.
- Great inter-annual variability.
- Significant flow.

Pluvial

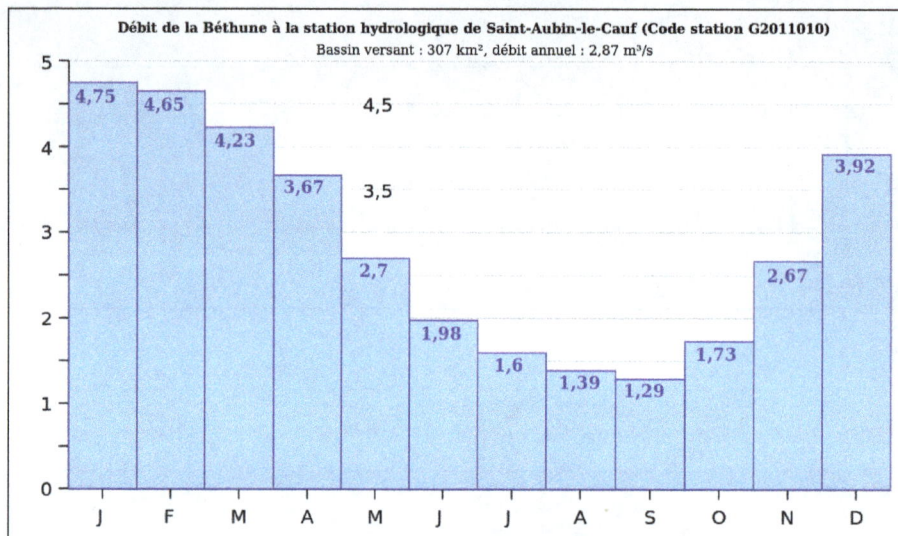

Pluvial-oceanic regime of the Béthune (France).

The pluvial regime is characterized by:

- High water in winter and spring.
- Low discharge in summer.
- Great inter-annual variability.
- Flow is generally rather weak.

It is typical of rivers at low to moderate altitude (500 to 1,000 metres or 1,600 to 3,300 feet).

Example: Seine.

Tropical Pluvial

The tropical pluvial regime is characterized by:

- Very low discharge in the cold season and abundant rainfall in the warm season.
- Minimum can reach very low values.
- Great variability of discharge during the year.
- Relatively regular from one year to another.

Mixed Régimes/Double Regimc

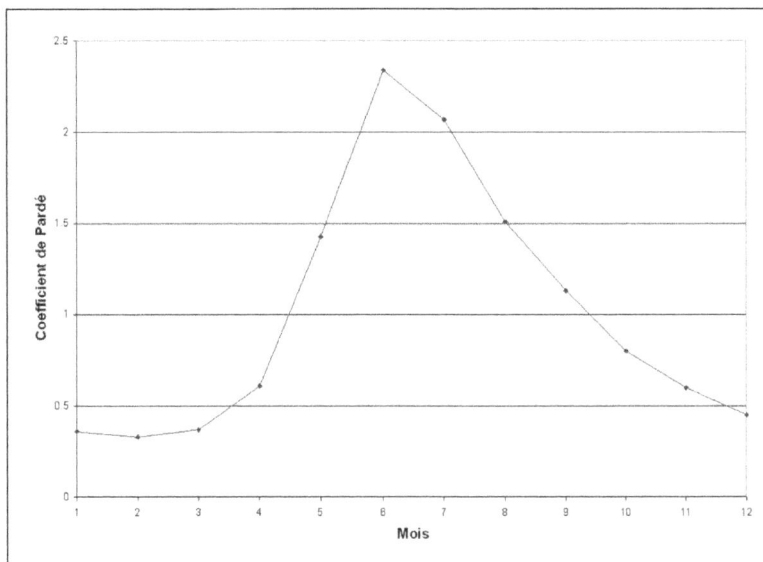

Nivo-glacial regime of the Albula (Switzerland)

Nivo-glacial

- Only one true maximum, which occurs in the late spring or the early summer (from May to July in the case of the Northern hemisphere).

- Relatively high diurnal variations during the hot season.

- Significant yearly variation, but less than in the snow regime.

- Significant flow.

Nivo-pluvial

- Two maximums, the first occurring in the spring and the other in autumn.

- A main low-water in October and a secondary low-water in January.

- Significant inter-annual variations.

Example: Issole [fr].

Pluvio-nival

- A period of rainfall in late autumn due to abundant rainfall, followed by a light increase due to snow melt in early spring.

- The single minimum occurs in autumn.

- Low amplitude.

Example: Mississippi.

Complex Regimes

The complex regime is characteristic of large rivers, the flow of which is diversely influenced by numerous tributaries from different altitudes, climates etc. The influences diminish extreme discharges and increase the regularity of the mean monthly discharge from upstream to downstream.

References

- Hydrograph: geology.com, Retrieved 11 February 2019

- Storm-hydrographs-and-river-discharge, river-profiles, geography: s-cool.co.uk, Retrieved 28 May 2019

- Luna B. Leopold; M. Gordon Wolman; John P. Miller. (1995). Fluvial processes in geomorphology. New York: Dover Publications. ISBN 0-486-68588-8

- Flood-hydrographs: internetgeography.net, Retrieved 20 April 2019

4

Water Purification Techniques

The process of removal of the biological contaminants, undesirable chemicals, suspended solids from water is known as water purification. Granular activated carbon adsorption, distillation, reverse osmosis, desalination, etc. are some of the techniques that fall under its domain. All the diverse applications of these water purification techniques have been carefully analyzed in this chapter.

Water Purification

Water purification is a process by which undesired chemical compounds, organic and inorganic materials, and biological contaminants are removed from water. That process also includes distillation (the conversion of a liquid into vapour to condense it back to liquid form) and deionization (ion removal through the extraction of dissolved salts). One major purpose of water purification is to provide clean drinking water. Water purification also meets the needs of medical, pharmacological, chemical, and industrial applications for clean and potable water. The purification procedure reduces the concentration of contaminants such as suspended particles, parasites, bacteria, algae, viruses, and fungi. Water purification takes place on scales from the large (e.g., for an entire city) to the small (e.g., for individual households).

Most communities rely on natural bodies of water as intake sources for water purification and for day-to-day use. In general, these resources can be classified as groundwater or surface water and commonly include underground aquifers, creeks, streams, rivers, and lakes. With recent technological advancements, oceans and saltwater seas have also been used as alternative water sources for drinking and domestic use.

Determining Water Quality

Historical evidence suggests that water treatment was recognized and practiced by ancient civilizations. Basic treatments for water purification have been documented in Greek and Sanskrit writings, and Egyptians used alum for precipitation as early as 1500 BCE.

In modern times, the quality to which water must be purified is typically set by government agencies. Whether set locally, nationally, or internationally, government standards typically set maximum concentrations of harmful contaminants that can be allowed in safe water. Since it is nearly impossible to examine water simply on the basis of appearance, multiple processes, such as physical, chemical, or biological analyses, have been developed to test contamination levels. Levels of organic and inorganic chemicals, such as chloride, copper, manganese, sulfates, and zinc, microbial pathogens, radioactive materials, and dissolved and suspended solids, as well as pH, odour, colour, and taste, are some of the common parameters analyzed to assess water quality and contamination levels.

Regular household methods such as boiling water or using an activated-carbon filter can remove some water contaminants. Although those methods are popular because they can be used widely and inexpensively, they often do not remove more dangerous contaminants. For example, natural spring water from artesian wells was historically considered clean for all practical purposes, but it came under scrutiny during the first decade of the 21st century because of worries over pesticides, fertilizers, and other chemicals from the surface entering wells. As a result, artesian wells were subjected to treatment and batteries of tests, including tests for the parasite Cryptosporidium.

Not all people have access to safe drinking water. According to a 2017 report by the United Nations (UN) World Health Organization (WHO), 2.1 billion people lack access to a safe and reliable drinking water supply at home. Eighty-eight percent of the four billion annual cases of diarrhea reported worldwide have been attributed to a lack of sanitary drinking water. Each year approximately 525,000 children under age five die from diarrhea, the second leading cause of death, and 1.7 million are sickened by diarrheal diseases caused by unsafe water, coupled with inadequate sanitation and hygiene.

Boiling

Boiling drinking water with fuel is the oldest and most commonly practiced household water treatment method. According to WHO, water needs to be heated until the appearance of the first big bubbles to ensure that it is pathogen free. Many organizations recommend boiling both for water treatment in developing countries and to provide safe drinking water in emergency situations throughout the world - though it is quite laborious and uses a lot of energy. Boiling only kills pathogens and does not remove turbidity or chemical pollution (e.g. arsenic) from drinking water. So prior to boiling, water can be purified by settling or filtration method.

Advantages

- Effectively kills most pathogens.

- Easy, simple and widely accepted method of disinfection (particularly in tea-consuming cultures).

- Biogas cooking stoves can be used for the cooking stove (e.g. biogas linked toilet).

Disadvantages

- Can be costly due to fuel consumptions.

- Use of traditional fuel (firewood, kerosene/gas) contributes to deforestation and indoor air pollution.

- Potential user taste objections.

- Risk of injuries (especially when children are around).

- Does not remove turbidity, chemicals, taste, smell, colour.

- Time consuming.

- Water needs to cool down before use unless for hot drinks.

Granular Activated Carbon Adsorption

Activated carbon is commonly used to adsorb natural organic compounds, taste and odor compounds, and synthetic organic chemicals in drinking water treatment. Adsorption is both the physical and chemical process of accumulating a substance at the interface between liquid and solids phases. Activated carbon is an effective adsorbent because it is a highly porous material and provides a large surface area to which contaminants may adsorb. The two main types of activated carbon used in water treatment applications are granular activated carbon (GAC) and powdered activated carbon (PAC).

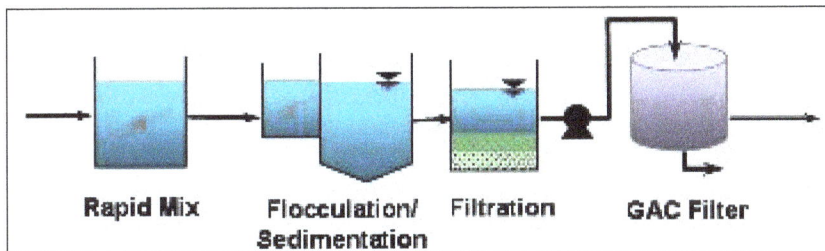

Post-filtration adsorption.

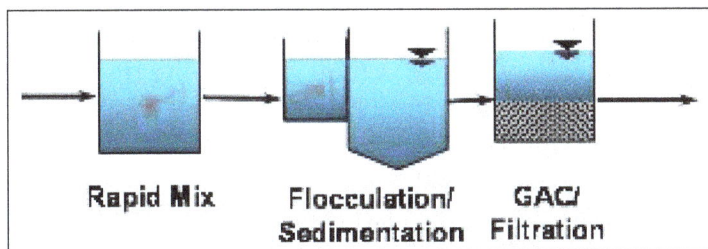

Filtration/adsorption.

GAC is made from organic materials with high carbon contents such as wood, lignite and coal. The primary characteristic that differentiates GAC to PAC is its particle size. GAC typically has a diameter ranging between 1.2 to 1.6 mm and an apparent density ranging between 25 and 31 lb/ft³), depending on the material used and manufacturing process. The bed density is about 10 percent less than the apparent density and is used to determine the amount of GAC required to fill a given size filter. The uniformity coefficient of GAC is quite large, typically about 1.9, to promote stratification after backwashing and minimize desorption and premature breakthrough that can result from mixing activated carbon particles with adsorbed compounds with activated carbon particles with smaller amounts of adsorbed compounds. Iodine and molasses numbers are typically used to characterize GAC. These numbers describe the quantity of small and large pore volumes in a sample of GAC. A minimum iodine number of 500 is specified for activated carbon by AWWA standards.

The two most common options for locating a GAC treatment unit in water treatment plants are: (1) post-filtration adsorption, where the GAC unit is located after the conventional filtration process (post-filter contactors or adsorbers); and (2) filtration-adsorption, in which some or all of the filter media in a granular media filter is replaced with GAC. Examples of these configurations are shown in figures 1 and 2, respectively.

In post-filtration applications, the GAC contactor receives the highest quality water and, thus, has as its only objective the removal of dissolved organic compounds. Backwashing of these adsorbers is usually unnecessary, unless excessive biological growth occurs. This option provides the most flexibility for handling GAC and for designing specific adsorption conditions by providing longer contact times than filter-adsorbers.

In addition to dissolved organics removal, the filter-adsorber configuration uses the GAC for turbidity and solids removal, and biological stabilization. Existing rapid sand filters can frequently be retrofitted for filtration-adsorption by replacing all or a portion of the granular media with GAC. Retrofitting existing high rate granular media filters can significantly reduce capital costs since no additional filter boxes, underdrains and backwashing systems may be required. However, filter-adsorbers have shorter filter run times and must be backwashed more frequently than post-filter adsorbers (filter-adsorber units are backwashed about as frequently as conventional high rate granular filters). In addition, filter-adsorbers may incur greater carbon losses because of increased backwashing and may cost more to operate because carbon usage is less effective.

Primary factors in determining the required GAC contactor volume are the (1) breakthrough, (2) empty bed contact time (EBCT), and (3) design flow rate. The breakthrough time is the time when the concentration of a contaminant in the effluent of the GAC unit exceeds the treatment requirement. As a rule of thumb, if the GAC effluent concentration is greater than the performance standard for over three consecutive days, the GAC is exhausted and must be replaced/regenerated. The EBCT is calculated as the empty bed volume divided by the flowrate through the carbon. Longer EBCTs can be achieved by increasing the bed volume or reducing the flow rate through the filter. The EBCT and the design flow rate define the amount of carbon to be contained in the adsorption units. A longer EBCT can delay breakthrough and reduce the GAC replacement/regeneration frequency. The carbon depth and adsorber volume can be determined once the optimum EBCT is established. Typical EBCTs for water treatment applications range between 5 to 25 minutes.

The surface loading rate for GAC filters is the flow rate through a given area of GAC filter bed and is expressed in units of gpm/ft2. Surface loading rates for GAC filters typically range between 2 to 10 gpm/ft2. High surface loading rates can be used when highly adsorbable compounds (such as SOCs) are targeted for removal. The surface loading rate is not important when mass transfer is controlled by the rate of adsorption as is the case for less-adsorbable compounds.

The carbon usage rate (CUR) determines the rate at which carbon will be exhausted and how often carbon must be replaced/regenerated. Carbon treatment effectiveness improves with increasing contact times. Deeper beds will increase the percentage of carbon that is exhausted at breakthrough. The optimum bed depth and volume are typically selected after carefully evaluating capital and operating costs associated with reactivation frequency and contactor construction costs.

GAC contactors can be configured as either (1) downflow fixed beds, (2) upflow fixed or expanded beds, or (3) pulsed beds; with single or multiple adsorbers operated in series or in parallel. In downflow fixed beds in series, each unit is connected in series with the first adsorber receiving the highest contaminant loading and the last unit receiving the lightest contaminant load. Carbon is removed for reactivation from the first unit, with the next adsorber becoming the lead unit. For downflow fixed beds in parallel, each unit receives the same flow and contaminant load. To maximize carbon usage, multiple contactors are frequently operated in parallel-staggered mode in which each contactor is at a different stage of carbon exhaustion. Since effluent from each contactor is blended, individual contactors can be operated beyond breakthrough such that the blended flow still meets the treatment goal. Upflow expanded beds permit removal of suspended solids by periodic bed expansion and allow using smaller carbon particles without significantly increasing head loss. In pulsed bed adsorbers, removal of spent carbon occurs from the bottom of the bed while fresh carbon is added at the top without system shutdown. A pulsed bed cannot be completely exhausted, which prevents contaminant breakthrough in the effluent.

Depending on the economics, facilities may have on-site or off-site regeneration systems or may waste spent carbon and replace it with new. Spent GAC must be disposed of recognizing that contaminants can be desorbed, which can potentially result in leaching of contaminants from the spent GAC when exposed to percolating water, contaminating soils or groundwater. Due to contamination concerns, spent GAC regeneration is typically favored over disposal. The three most common GAC regeneration methods are steam, thermal and chemical; of which thermal regeneration is the most common method used. Available thermal regeneration technologies used to remove adsorbed organics from activated carbon include: (1) electric infrared ovens, (2) fluidized bed furnaces, (3) multiple hearth furnaces, and (4) rotary kilns.

Distillation

Distillation is one of the oldest methods of water treatment and is still in use today, though not commonly as a home treatment method. It can effectively remove many contaminants from drinking water, including bacteria, inorganic and many organic compounds.

Your first step toward solving a suspected water quality problem is having your water analyzed by the local health department or a reputable laboratory. A water analysis not only verifies if a water quality problem exists, but is essential to determine the most appropriate solution to the problem. State or local health officials can interpret water analysis results. Some laboratories may also provide this service.

Home water treatment should be considered only a temporary solution. The best solutions to a contaminated drinking water problem are to stop the practices causing the contamination or change water sources.

Distillation relies on evaporation to purify water. Contaminated water is heated to form steam. Inorganic compounds and large non-volatile organic molecules do not evaporate with the water and are left behind. The steam then cools and condenses to form purified water.

Distillation effectively removes inorganic compounds such as metals (lead), nitrate, and other nuisance particles such as iron and hardness from a con- taminated water supply. The boiling process also kills microorganisms such as bacteria and some viruses. Distillation removes oxygen and some trace metals from water. For this reason some people claim distilled water tastes flat.

Distillation's effectiveness in removing organic compounds varies, depending on such chemical characteristics of the organic compound as solubility and boiling point. Organic compounds with boiling points lower than the boiling point of water (ex. benzene and toluene) vaporize along with the water. These harmful compounds will recontaminate the purified product if not removed prior to condensation.

Distillation Units

Distillation units, or stills, generally consist of a boiling chamber, where the water enters, is heated and vaporized; condensing coils or chamber, where the water is cooled and converted back to liquid water; and a storage tank for purified water. Figure shows the parts and process of a distiller.

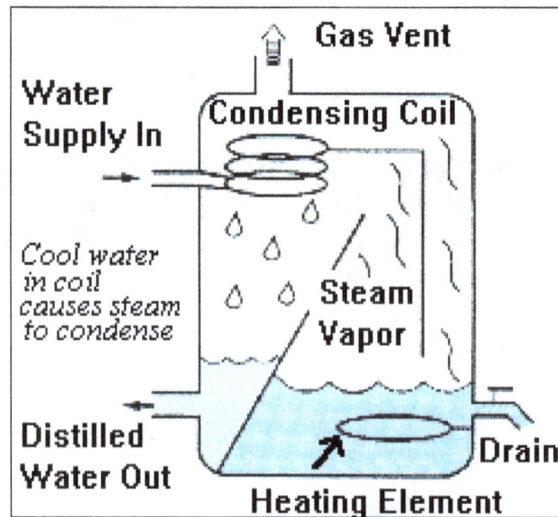

Parts and process of a distiller.

Distillation units are usually installed as point-of-use (POU) systems. They are generally placed at the kitchen faucet and used to purify water for drinking and cooking only. Size varies, depending on the amount of purified water they produce. Production rates range from 3 to 11 gallons per day. Home stills can be located on the counter or floor, or attached to the wall. Models can be fully or partially automated, or manual.

Some units have columns or volatile gas vents to eliminate organic chemicals with boiling points lower than water, thus ensuring uncontaminated water.

Operation, Maintenance and Cost

As with all home water treatment systems, distillation units require some level of regular maintenance to keep the unit operating properly. Uneva- porated pollutants left in the boiling chamber need to be regularly flushed to the septic or sewer system. Even with regular removal of the residual water containing unevaporated pollutants, a calcium and magnesium scale will collect at the bottom of the boiling chamber. The scale eventually needs to be removed, usually by hand scrubbing or by using acid.

Heating water to form steam requires energy. Which means the operating costs for distillation units are generally higher than those for other home water treatments. The production of heat from home distillation units can be an advantage in winter, but a disadvantage in summer.

Distillation units are generally expensive, ranging from $300 to $1200. Port- able units can be purchased for less than $200.

Certification and Validation

Certification of treatment products is available from independent testing laboratories, such as the National Sanitation Foundation (NSF). Results from NSF tests provide good measures of the effectiveness of devices designed to treat water for both esthetic and health reasons. The Water Quality Associa- tion (WQA) is a self-governing body of manufacturers and distributors. WQA offers voluntary validation programs to its members. Validation is less stringent than certification. Certification or validation does not ensure effective treatment; all systems must be designed for each particular situa- tion and maintained properly.

Solar Distillation Water Treatment

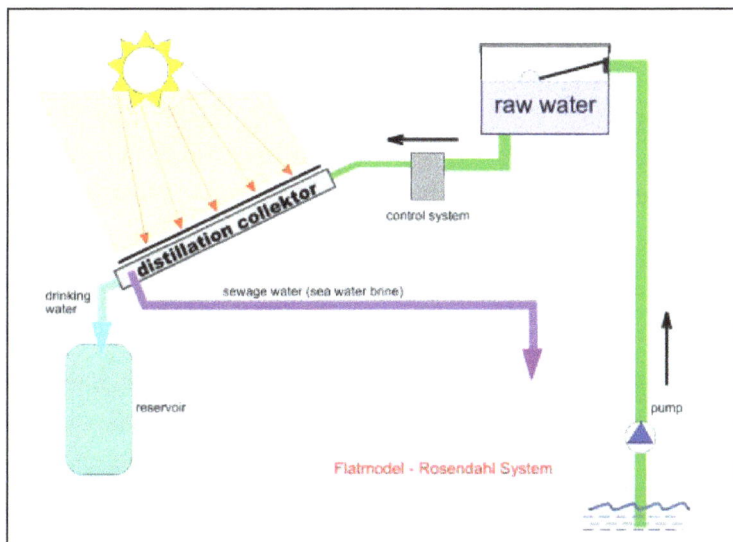

Solar Distillation.

For heating water solar energy can be harnessed by the use of a system of mirrors following the path of the sun to focus the sunlight on sheets of water in a solar still. In solar stills, salt water is filled in basins with glass panes as roof at an angle of 10 to 18; tightly sealed to the holding frame and the joints between the still cover and the vertical walls perfectly tight. Upon heating by solar heat the water molecules are converted into steam leaving behind the salts at the basin. The steam is condensed at the bottom of the glass roof of the still and water slides along the slopes of the glass panel and collected in the collecting troughs. The collecting troughs at the foot of the glass pane cover of the still must be constructed in such a way that water will drain freely to the pipe which carries the distillate to the fresh water tank.

Multi-stage Flash Distillation Water Treatment

In multi stage flash (MSF) distillation, the feed water is heated and the pressure is lowered, so the water flashes into steam. This process constitutes one stage of a number of stages in series, each of which is at a lower pressure.

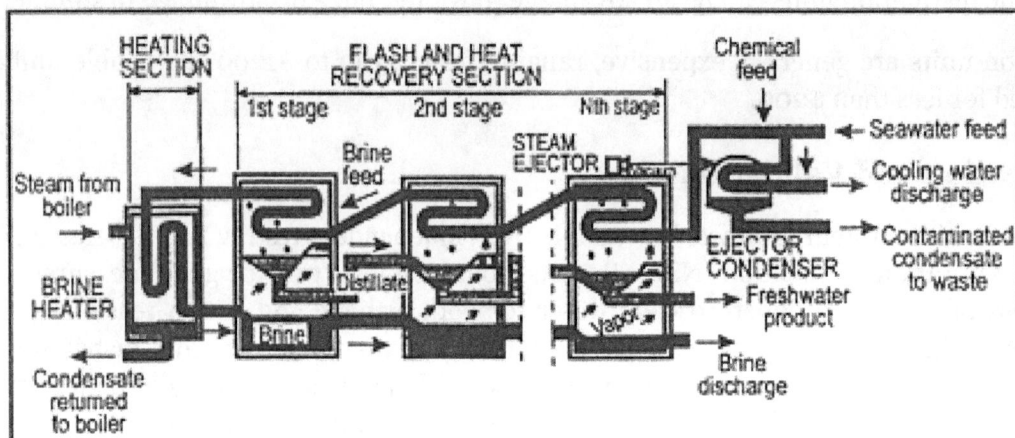

Multi Stage Flash Distillation.

Distilled Water Removes Minerals and Contaminants

Distillation will not remove all the chemicals but removes soluble minerals (i.e., calcium, magnesium, and phosphorous) and dangerous heavy metals like lead, arsenic, and mercury. Some of the chemicals of concern produce hazardous compounds during the heating process. The vaporization process strips salt, metals, and biological threats. Stripping of minerals will not be harmful to the human body system as stated by World Health Organization (WHO).

Reverse Osmosis

Reverse osmosis (RO) is a water purification process that uses a partially permeable membrane to remove ions, unwanted molecules and larger particles from drinking water. In reverse osmosis, an applied pressure is used to overcome osmotic pressure, a colligative property, that is driven by chemical potential differences of the solvent, a thermodynamic parameter. Reverse osmosis can remove many types of dissolved and suspended chemical species as well as biological ones (principally bacteria) from water, and is used in both industrial processes and the production of potable water. The result is that the solute is retained on the pressurized side of the membrane and the pure solvent is allowed to pass to the other side. To be "selective", this membrane should not allow large molecules or ions through the pores (holes), but should allow smaller components of the solution (such as solvent molecules, i.e., water, H_2O) to pass freely.

In the normal osmosis process, the solvent naturally moves from an area of low solute concentration (high water potential), through a membrane, to an area of high solute concentration (low water potential). The driving force for the movement of the solvent is the reduction in the free energy of the system when the difference in solvent concentration on either side of a membrane is reduced, generating osmotic pressure due to the solvent moving into the more concentrated solution. Applying an external pressure to reverse the natural flow of pure solvent, thus, is reverse osmosis. The process is similar to other membrane technology applications.

Reverse osmosis differs from filtration in that the mechanism of fluid flow is by osmosis across a membrane. The predominant removal mechanism in membrane filtration is straining, or size exclusion, where the pores are 0.01 micrometers or larger, so the process can theoretically achieve perfect efficiency regardless of parameters such as the solution's pressure and concentration. Reverse osmosis instead involves solvent diffusion across a membrane that is either nonporous or uses nanofiltration with pores 0.001 micrometers in size. The predominant removal mechanism is from differences in solubility or diffusivity, and the process is dependent on pressure, solute concentration, and other conditions. Reverse osmosis is most commonly known for its use in drinking water purification from seawater, removing the salt and other effluent materials from the water molecules.

Fresh Water Applications

Drinking Water Purification

Around the world, household drinking water purification systems, including a reverse osmosis step, are commonly used for improving water for drinking and cooking.

Such systems typically include a number of steps:

- A sediment filter to trap particles, including rust and calcium carbonate.

- Optionally, a second sediment filter with smaller pores.

- An activated carbon filter to trap organic chemicals and chlorine, which will attack and degrade a thin film composite membrane.

- A reverse osmosis filter, which is a thin film composite membrane.

- Optionally, a second carbon filter to capture those chemicals not removed by the reverse osmosis membrane.

- Optionally an ultraviolet lamp for sterilizing any microbes that may escape filtering by the reverse osmosis membrane.

The latest developments in the sphere include nano materials and membranes.

In some systems, the carbon prefilter is omitted, and a cellulose triacetate membrane is used. CTA (cellulose triacetate) is a paper by-product membrane bonded to a synthetic layer and is made to allow contact with chlorine in the water. These require a small amount of chlorine in the water source to prevent bacteria from forming on it. The typical rejection rate for CTA membranes is 85–95%.

The cellulose triacetate membrane is prone to rotting unless protected by chlorinated water, while the thin film composite membrane is prone to breaking down under the influence of chlorine. A thin film composite (TFC) membrane is made of synthetic material, and requires chlorine to be removed before the water enters the membrane. To protect the TFC membrane elements from chlorine damage, carbon filters are used as pre-treatment in all residential reverse osmosis systems. TFC membranes have a higher rejection rate of 95–98% and a longer life than CTA membranes.

Portable reverse osmosis water processors are sold for personal water purification in various locations. To work effectively, the water feeding to these units should be under some pressure (280 kPa (40 psi) or greater is the norm). Portable reverse osmosis water processors can be used by people who live in rural areas without clean water, far away from the city's water pipes. Rural people filter river or ocean water themselves, as the device is easy to use (saline water may need special membranes). Some travelers on long boating, fishing, or island camping trips, or in countries where the local water supply is polluted or substandard, use reverse osmosis water processors coupled with one or more ultraviolet sterilizers.

In the production of bottled mineral water, the water passes through a reverse osmosis water processor to remove pollutants and microorganisms. In European countries, though, such processing of natural mineral water (as defined by a European directive) is not allowed under European law. In practice, a fraction of the living bacteria can and do pass through reverse osmosis membranes through minor imperfections, or bypass the membrane entirely through tiny leaks in surrounding seals. Thus, complete reverse osmosis systems may include additional water treatment stages that use ultraviolet light or ozone to prevent microbiological contamination.

Membrane pore sizes can vary from 0.1 to 5,000 nm depending on filter type. Particle filtration removes particles of 1 μm or larger. Microfiltration removes particles of 50 nm or larger. Ultrafiltration removes particles of roughly 3 nm or larger. Nanofiltration removes particles of 1 nm or larger. Reverse osmosis is in the final category of membrane filtration, hyperfiltration, and removes particles larger than 0.1 nm.

Decentralized use: Solar-powered Reverse Osmosis

A solar-powered desalination unit produces potable water from saline water by using a photovoltaic system that converts solar power into the required energy for reverse osmosis. Due to the extensive availability of sunlight across different geographies, solar-powered reverse osmosis lends itself well to drinking water purification in remote settings lacking an electricity grid. Moreover, Solar energy overcomes the usually high-energy operating costs as well as greenhouse emissions of conventional reverse osmosis systems, making it a sustainable freshwater solution compatible to developing contexts. For example, a solar-powered desalination unit designed for remote communities has been successfully tested in the Northern Territory of Australia.

While the intermittent nature of sunlight and its variable intensity throughout the day makes PV efficiency prediction difficult and desalination during night time challenging, several solutions exist. For example, batteries, which provide the energy required for desalination in non-sunlight hours can be used to store solar energy in daytime. Apart from the use of conventional batteries, alternative methods for solar energy storage exist. For example, thermal energy storage systems solve this storage problem and ensure constant performance even during non-sunlight hours and cloudy days, improving overall efficiency.

Military use: The Reverse Osmosis Water Purification Unit

A reverse osmosis water purification unit (ROWPU) is a portable, self-contained water treatment plant. Designed for military use, it can provide potable water from nearly any water source. There are many models in use by the United States armed forces and the Canadian Forces. Some models are containerized, some are trailers, and some are vehicles unto themselves.

Each branch of the United States armed forces has their own series of reverse osmosis water purification unit models, but they are all similar. The water is pumped from its raw source into the reverse osmosis water purification unit module, where it is treated with a polymer to initiate coagulation. Next, it is run through a multi-media filter where it undergoes primary treatment by removing turbidity. It is then pumped through a cartridge filter which is usually spiral-wound cotton. This process clarifies the water of any particles larger than 5 μm and eliminates almost all turbidity.

The clarified water is then fed through a high-pressure piston pump into a series of vessels where it is subject to reverse osmosis. The product water is free of 90.00–99.98% of the raw water's total dissolved solids and by military standards, should have no more than 1000–1500 parts per million by measure of electrical conductivity. It is then disinfected with chlorine and stored for later use.

Within the United States Marine Corps, the reverse osmosis water purification unit has been replaced by both the Lightweight Water Purification System and Tactical Water Purification Systems. The Lightweight Water Purification Systems can be transported by Humvee and filter 470 litres (120 US gal) per hour. The Tactical Water Purification Systems can be carried on a Medium Tactical Vehicle Replacement truck, and can filter 4,500 to 5,700 litres (1,200 to 1,500 US gal) per hour.

Water and Wastewater Purification

Rain water collected from storm drains is purified with reverse osmosis water processors and used for landscape irrigation and industrial cooling in Los Angeles and other cities, as a solution to the problem of water shortages.

In industry, reverse osmosis removes minerals from boiler water at power plants. The water is distilled multiple times. It must be as pure as possible so it does not leave deposits on the machinery or cause corrosion. The deposits inside or outside the boiler tubes may result in under-performance of the boiler, reducing its efficiency and resulting in poor steam production, hence poor power production at the turbine.

It is also used to clean effluent and brackish groundwater. The effluent in larger volumes (more than 500 m³/day) should be treated in an effluent treatment plant first, and then the clear effluent is subjected to reverse osmosis system. Treatment cost is reduced significantly and membrane life of the reverse osmosis system is increased. The process of reverse osmosis can be used for the production of deionized water.

Reverse osmosis process for water purification does not require thermal energy. Flow-through reverse osmosis systems can be regulated by high-pressure pumps. The recovery of purified water depends upon various factors, including membrane sizes, membrane pore size, temperature, operating pressure, and membrane surface area.

In 2002, Singapore announced that a process named NEWater would be a significant part of its future water plans. It involves using reverse osmosis to treat domestic wastewater before discharging the NEWater back into the reservoirs.

Food Industry

In addition to desalination, reverse osmosis is a more economical operation for concentrating food liquids (such as fruit juices) than conventional heat-treatment processes. Research has been done on concentration of orange juice and tomato juice. Its advantages include a lower operating cost and the ability to avoid heat-treatment processes, which makes it suitable for heat-sensitive substances such as the protein and enzymes found in most food products.

Reverse osmosis is extensively used in the dairy industry for the production of whey protein powders and for the concentration of milk to reduce shipping costs. In whey applications, the whey (liquid remaining after cheese manufacture) is concentrated with reverse osmosis from 6% total solids to 10–20% total solids before ultrafiltration processing. The ultrafiltration retentate can then be used to make various whey powders, including whey protein isolate. Additionally, the ultrafiltration permeate, which contains lactose, is concentrated by reverse osmosis from 5% total solids to 18–22% total solids to reduce crystallization and drying costs of the lactose powder.

Although use of the process was once avoided in the wine industry, it is now widely understood and used. An estimated 60 reverse osmosis machines were in use in Bordeaux, France, in 2002. Known users include many of the elite-classed growths (Kramer) such as Château Léoville-Las Cases in Bordeaux.

Maple Syrup Production

In 1946, some maple syrup producers started using reverse osmosis to remove water from sap before the sap is boiled down to syrup. The use of reverse osmosis allows about 75–90% of the water to be removed from the sap, reducing energy consumption and exposure of the syrup to high temperatures. Microbial contamination and degradation of the membranes must be monitored.

Hydrogen Production

For small-scale hydrogen production, reverse osmosis is sometimes used to prevent formation of mineral deposits on the surface of electrodes.

Aquariums

Many reef aquarium keepers use reverse osmosis systems for their artificial mixture of seawater. Ordinary tap water can contain excessive chlorine, chloramines, copper, nitrates, nitrites, phosphates, silicates, or many other chemicals detrimental to the sensitive organisms in a reef environment. Contaminants such as nitrogen compounds and phosphates can lead to excessive and unwanted algae growth. An effective combination of both reverse osmosis and deionization is the most popular among reef aquarium keepers, and is preferred above other water purification processes due to the low cost of ownership and minimal operating costs. Where chlorine and chloramines are found in the water, carbon filtration is needed before the membrane, as the common residential membrane used by reef keepers does not cope with these compounds.

Freshwater aquarists also use reverse osmosis systems to duplicate the very soft waters found in many tropical water bodies. Whilst many tropical fish can survive in suitably treated tap water, breeding can be impossible. Many aquatic shops sell containers of reverse osmosis water for this purpose.

Window Cleaning

An increasingly popular method of cleaning windows is the so-called "water-fed pole" system. Instead of washing the windows with detergent in the conventional way, they are scrubbed with highly purified water, typically containing less than 10 ppm dissolved solids, using a brush on the end of a long pole which is wielded from ground level. Reverse osmosis is commonly used to purify the water.

Landfill Leachate Purification

Treatment with reverse osmosis is limited, resulting in low recoveries on high concentration (measured with electrical conductivity) and fouling of the RO membranes. Reverse osmosis applicability is limited by conductivity, organics, and scaling inorganic elements such as $CaSO_4$, Si, Fe and Ba. Low organic scaling can use two different technologies, one is using spiral wound membrane type of module, and for high organic scaling, high conductivity and higher pressure (up to 90 bars) disc tube modules with reverse-osmosis membranes can be used. Disc tube modules were redesigned for landfill leachate purification, that is usually contaminated with high levels of organic material. Due to the cross-flow with high velocity it is given a flow booster pump, that is recirculating the flow over the same membrane surface between 1.5 and 3 times before it is released as a concentrate. High velocity is also good against membrane scaling and allows successful membrane cleaning.

Power Consumption for a Disc Tube Module System

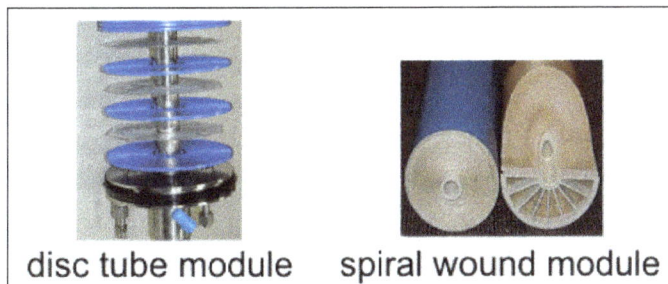

disc tube module   spiral wound module

Disc tube module with RO membrane cushion
and Spiral wound module with RO membrane.

Energy consumption per m³ leachate			
Name of module	1-stage up to 75 bar	2-stage up to 75 bar	3-stage up to 120 bar
Disc tube module	6.1 – 8.1 kWh/m³	8.1 – 9.8 kWh/m³	11.2 – 14.3 kWh/m³

Desalination

Areas that have either no or limited surface water or groundwater may choose to desalinate. Reverse osmosis is an increasingly common method of desalination, because of its relatively low energy consumption.

In recent years, energy consumption has dropped to around 3 kWh/m³, with the development of more efficient energy recovery devices and improved membrane materials. According to the International Desalination Association, for 2011, reverse osmosis was used in 66% of installed desalination capacity (0.0445 of 0.0674 km³/day), and nearly all new plants. Other plants mainly use thermal distillation methods: multiple-effect distillation and multi-stage flash.

Sea-water reverse-osmosis (SWRO) desalination, a membrane process, has been commercially used since the early 1970s. Its first practical use was demonstrated by Sidney Loeb from University of California at Los Angeles in Coalinga, California, and Srinivasa Sourirajan of National Research Council, Canada. Because no heating or phase changes are needed, energy requirements are low, around 3 kWh/m³, in comparison to other processes of desalination, but are still much higher than those required for other forms of water supply, including reverse osmosis treatment of wastewater, at 0.1 to 1 kWh/m³. Up to 50% of the seawater input can be recovered as fresh water, though lower recoveries may reduce membrane fouling and energy consumption.

Brackish water reverse osmosis refers to desalination of water with a lower salt content than sea water, usually from river estuaries or saline wells. The process is substantially the same as sea water reverse osmosis, but requires lower pressures and therefore less energy. Up to 80% of the feed water input can be recovered as fresh water, depending on feed salinity.

The Ashkelon sea water reverse osmosis desalination plant in Israel is the largest in the world. The project was developed as a build-operate-transfer by a consortium of three international companies: Veolia water, IDE Technologies, and Elran.

The typical single-pass sea water reverse osmosis system consists of:

- Intake,
- Pretreatment,
- High-pressure pump (if not combined with energy recovery),
- Membrane assembly,
- Energy recovery (if used),
- Remineralisation and pH adjustment,
- Disinfection,
- Alarm/control panel.

Pretreatment

Pretreatment is important when working with reverse osmosis and nanofiltration membranes due to the nature of their spiral-wound design. The material is engineered in such a fashion as to allow only one-way flow through the system. As such, the spiral-wound design does not allow for backpulsing with water or air agitation to scour its surface and remove solids. Since accumulated material cannot be removed from the membrane surface systems, they are highly susceptible to fouling (loss of production capacity). Therefore, pretreatment is a necessity for any reverse osmosis or nanofiltration system. Pretreatment in sea water reverse osmosis systems has four major components:

- Screening of solids: Solids within the water must be removed and the water treated to prevent fouling of the membranes by fine-particle or biological growth, and reduce the risk of damage to high-pressure pump components.

- Cartridge filtration: Generally, string-wound polypropylene filters are used to remove particles of 1–5 µm diameter.

- Dosing: Oxidizing biocides, such as chlorine, are added to kill bacteria, followed by bisulfite dosing to deactivate the chlorine, which can destroy a thin-film composite membrane. There are also biofouling inhibitors, which do not kill bacteria, but simply prevent them from growing slime on the membrane surface and plant walls.

- Prefiltration pH adjustment: If the pH, hardness and the alkalinity in the feedwater result in a scaling tendency when they are concentrated in the reject stream, acid is dosed to maintain carbonates in their soluble carbonic acid form.

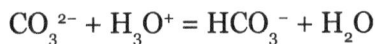

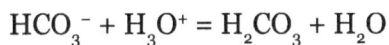

$$CO_3^{2-} + H_3O^+ = HCO_3^- + H_2O$$

$$HCO_3^- + H_3O^+ = H_2CO_3 + H_2O$$

- Carbonic acid cannot combine with calcium to form calcium carbonate scale. Calcium carbonate scaling tendency is estimated using the Langelier saturation index. Adding too much sulfuric acid to control carbonate scales may result in calcium sulfate, barium sulfate, or strontium sulfate scale formation on the reverse osmosis membrane.

- Prefiltration antiscalants: Scale inhibitors (also known as antiscalants) prevent formation of all scales compared to acid, which can only prevent formation of calcium carbonate and calcium phosphate scales. In addition to inhibiting carbonate and phosphate scales, antiscalants inhibit sulfate and fluoride scales and disperse colloids and metal oxides. Despite claims that antiscalants can inhibit silica formation, no concrete evidence proves that silica polymerization can be inhibited by antiscalants. Antiscalants can control acid-soluble scales at a fraction of the dosage required to control the same scale using sulfuric acid.

- Some small-scale desalination units use 'beach wells'; they are usually drilled on the seashore in close vicinity to the ocean. These intake facilities are relatively simple to build and the seawater they collect is pretreated via slow filtration through the subsurface sand/seabed formations in the area of source water extraction. Raw seawater collected using beach wells is often of better quality in terms of solids, silt, oil and grease, natural organic contamination and aquatic microorganisms, compared to open seawater intakes. Sometimes, beach intakes may also yield source water of lower salinity.

High Pressure Pump

The high pressure pump supplies the pressure needed to push water through the membrane, even as the membrane rejects the passage of salt through it. Typical pressures for brackish water range from 1.6 to 2.6 MPa (225 to 376 psi). In the case of seawater, they range from 5.5 to 8 MPa (800 to 1,180 psi). This requires a large amount of energy. Where energy recovery is used, part of the high pressure pump's work is done by the energy recovery device, reducing the system energy inputs.

Membrane Assembly

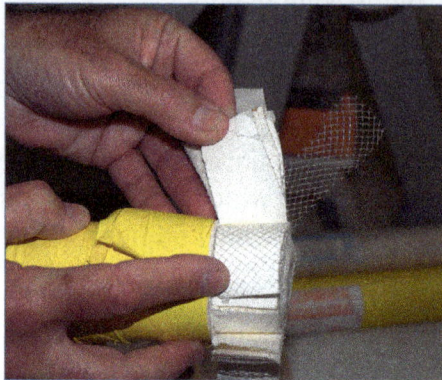

The layers of a membrane.

The membrane assembly consists of a pressure vessel with a membrane that allows feedwater to be pressed against it. The membrane must be strong enough to withstand whatever pressure is applied against it. Reverse-osmosis membranes are made in a variety of configurations, with the two most common configurations being spiral-wound and hollow-fiber.

Only a part of the saline feed water pumped into the membrane assembly passes through the membrane with the salt removed. The remaining "concentrate" flow passes along the saline side of the membrane to flush away the concentrated salt solution. The percentage of desalinated water produced versus the saline water feed flow is known as the "recovery ratio". This varies with the salinity of the feed water and the system design parameters: typically 20% for small seawater systems, 40% – 50% for larger seawater systems, and 80% – 85% for brackish water. The concentrate flow is at typically only 3 bar / 50 psi less than the feed pressure, and thus still carries much of the high-pressure pump input energy.

The desalinated water purity is a function of the feed water salinity, membrane selection and recovery ratio. To achieve higher purity a second pass can be added which generally requires re-pumping. Purity expressed as total dissolved solids typically varies from 100 to 400 parts per million (ppm or mg/litre)on a seawater feed. A level of 500 ppm is generally accepted as the upper limit for drinking water, while the US Food and Drug Administration classifies mineral water as water containing at least 250 ppm.

Energy Recovery

Energy recovery can reduce energy consumption by 50% or more. Much of the high pressure pump input energy can be recovered from the concentrate flow, and the increasing efficiency of energy recovery devices has greatly reduced the energy needs of reverse osmosis desalination. Devices used, in order of invention, are:

- Turbine or Pelton wheel: a water turbine driven by the concentrate flow, connected to the high pressure pump drive shaft to provide part of its input power. Positive displacement axial piston motors have also been used in place of turbines on smaller systems.

- Turbocharger: a water turbine driven by the concentrate flow, directly connected to a centrifugal pump which boosts the high pressure pump output pressure, reducing the pressure needed from the high pressure pump and thereby its energy input, similar in construction principle to car engine turbochargers.

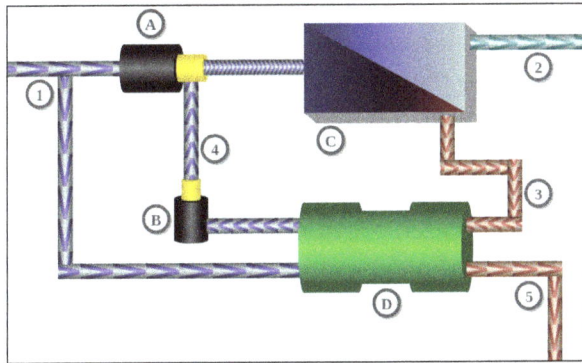

Schematics of a reverse osmosis desalination system using a pressure exchanger. 1: Sea water inflow, 2: Fresh water flow (40%), 3: Concentrate flow (60%), 4: Sea water flow (60%), 5: Concentrate (drain), A: Pump flow (40%), B: Circulation pump, C: Osmosis unit with membrane, D: Pressure exchanger.

- Pressure exchanger: using the pressurized concentrate flow, in direct contact or via a piston, to pressurize part of the membrane feed flow to near concentrate flow pressure. A boost pump then raises this pressure by typically 3 bar / 50 psi to the membrane feed pressure. This reduces flow needed from the high-pressure pump by an amount equal to the concentrate flow, typically 60%, and thereby its energy input. These are widely used on larger low-energy systems. They are capable of 3 kWh/m³ or less energy consumption.

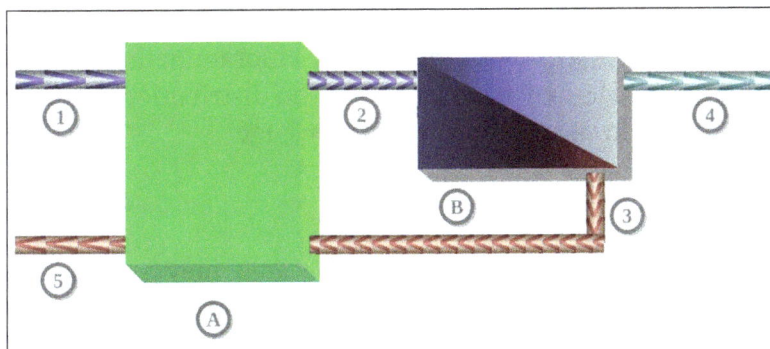

Schematic of a reverse osmosis desalination system using an energy recovery pump. *1*: Sea water inflow (100%, 1 bar), 2: Sea water flow (100%, 50 bar), 3: Concentrate flow (60%, 48 bar), 4: Fresh water flow (40%, 1 bar), 5: Concentrate to drain (60%,1 bar), A: *Pressure recovery pump*, B: *Osmosis unit with membrane*.

- Energy-recovery pump: a reciprocating piston pump having the pressurized concentrate flow applied to one side of each piston to help drive the membrane feed flow from the opposite side. These are the simplest energy recovery devices to apply, combining the high pressure pump and energy recovery in a single self-regulating unit. These are widely used on smaller low-energy systems. They are capable of 3 kWh/m³ or less energy consumption.

- Batch operation: Reverse-osmosis systems run with a fixed volume of fluid (thermodynamically a closed system) do not suffer from wasted energy in the brine stream, as the energy to pressurize a virtually incompressible fluid (water) is negligible. Such systems have the potential to reach second-law efficiencies of 60%.

Remineralisation and pH Adjustment

The desalinated water is stabilized to protect downstream pipelines and storage, usually by adding lime or caustic soda to prevent corrosion of concrete-lined surfaces. Liming material is used to adjust pH between 6.8 and 8.1 to meet the potable water specifications, primarily for effective disinfection and for corrosion control. Remineralisation may be needed to replace minerals removed from the water by desalination. Although this process has proved to be costly and not very convenient if it is intended to meet mineral demand by humans and plants. The very same mineral demand that freshwater sources provided previously. For instance water from Israel's national water carrier typically contains dissolved magnesium levels of 20 to 25 mg/liter, while water from the Ashkelon plant has no magnesium. After farmers used this water, magnesium-deficiency symptoms appeared in crops, including tomatoes, basil, and flowers, and had to be remedied by fertilization. Current Israeli drinking water standards set a minimum calcium level of 20 mg/liter. The postdesalination treatment in the Ashkelon plant uses sulfuric acid to dissolve calcite (limestone), resulting in calcium concentration of 40 to 46 mg/liter. This is still lower than the 45 to 60 mg/liter found in typical Israeli fresh water.

Disinfection

Post-treatment consists of preparing the water for distribution after filtration. Reverse osmosis is an effective barrier to pathogens, but post-treatment provides secondary protection against compromised membranes and downstream problems. Disinfection by means of ultra violet (UV) lamps (sometimes called germicidal or bactericidal) may be employed to sterilize pathogens which bypassed the reverse-osmosis process. Chlorination or chloramination (chlorine and ammonia) protects against pathogens which may have lodged in the distribution system downstream, such as from new construction, backwash, compromised pipes, etc.

Disadvantages

Household reverse-osmosis units use a lot of water because they have low back pressure. As a result, they recover only 5 to 15% of the water entering the system. The remainder is discharged as waste water. Because waste water carries with it the rejected contaminants, methods to recover this water are not practical for household systems. Wastewater is typically connected to the house drains and will add to the load on the household septic system. A reverse-osmosis unit delivering 19 L of treated water per day may discharge between 75–340 L of waste water daily. This has a disastrous consequence for mega cities like Delhi where large-scale use of household R.O. devices has increased the total water demand of the already water parched National Capital Territory of India.

Large-scale industrial/municipal systems recover typically 75% to 80% of the feed water, or as high as 90%, because they can generate the high pressure needed for higher recovery reverse osmosis filtration. On the other hand, as recovery of wastewater increases in commercial operations, effective contaminant removal rates tend to become reduced, as evidenced by product water total dissolved solids levels.

Reverse osmosis per its construction removes both harmful contaminants present in the water, as well as some desirable minerals. Modern studies on this matter have been quite shallow, citing lack of funding and interest in such study, as re-mineralization on the treatment plants today is done to prevent pipeline corrosion without going into human health aspect. They do, however link to older, more thorough studies that at one hand show some relation between long-term health effects and consumption of water low on calcium and magnesium, on the other confess that none of these older studies comply to modern standards of research.

Waste-stream Considerations

Depending upon the desired product, either the solvent or solute stream of reverse osmosis will be waste. For food concentration applications, the concentrated solute stream is the product and the solvent stream is waste. For water treatment applications, the solvent stream is purified water and the solute stream is concentrated waste. The solvent waste stream from food processing may be used as reclaimed water, but there may be fewer options for disposal of a concentrated waste solute stream. Ships may use marine dumping and coastal desalination plants typically use marine outfalls. Landlocked reverse osmosis plants may require evaporation ponds or injection wells to avoid polluting groundwater or surface runoff.

New Developments

Since the 1970s, prefiltration of high-fouling waters with another larger-pore membrane, with less hydraulic energy requirement, has been evaluated and sometimes used. However, this means that the water passes through two membranes and is often repressurized, which requires more energy to be put into the system, and thus increases the cost.

Other recent developmental work has focused on integrating reverse osmosis with electrodialysis to improve recovery of valuable deionized products, or to minimize the volume of concentrate requiring discharge or disposal. In the production of drinking water, the latest developments include nanoscale and graphene membranes.

The world's largest RO desalination plant was built in Sorek, Israel, in 2013. It has an output of 624,000 m^3 a day. It is also the cheapest and will sell water to the authorities for US$0.58/$m^3$.

Direct Contact Membrane Distillation

Membrane Distillation (MD) is a thermally driven transport process that uses hydrophobic membranes. The driving force in the method is the vapor pressure difference between the two sides of the membrane pores, allowing for mass and heat transfer of the volatile solution components (e.g. water). The simplicity of MD along with the fact that it can use waste heat and/or alternative energy sources, such as solar and geothermal energy, enables MD to be combined with other processes in integrated systems, making it a promising separation technique.

Advantages (+)	Disadvantages (-)
Treating >70,000 ppm (RO limit), No concentration limit	Temperature Concentration Polarization
No applied Pressure → ↓electrical Energy & ↓Fouling	Pore Wetting
↑Simplicity	Relative ↑thermal Energy demand
↓↓Chemicals	↑ Module cost
Not affected much by Concentration Polarization	↓ Flux
100% theoretical rejection of non volatile components	-

Energy Consumption; 47.41 KWh/m^3 = 2.03 KWh/m^3 Electrical + 45.38 KWh/m3 Thermal.

Applications

- Treatment of >70,000 ppm saline solution (Brine);

- Removal of volatile components (e.g. Ammonia);

- Water purification in the pharmaceutical, chemical and textile industries;

- Food & Beverages.

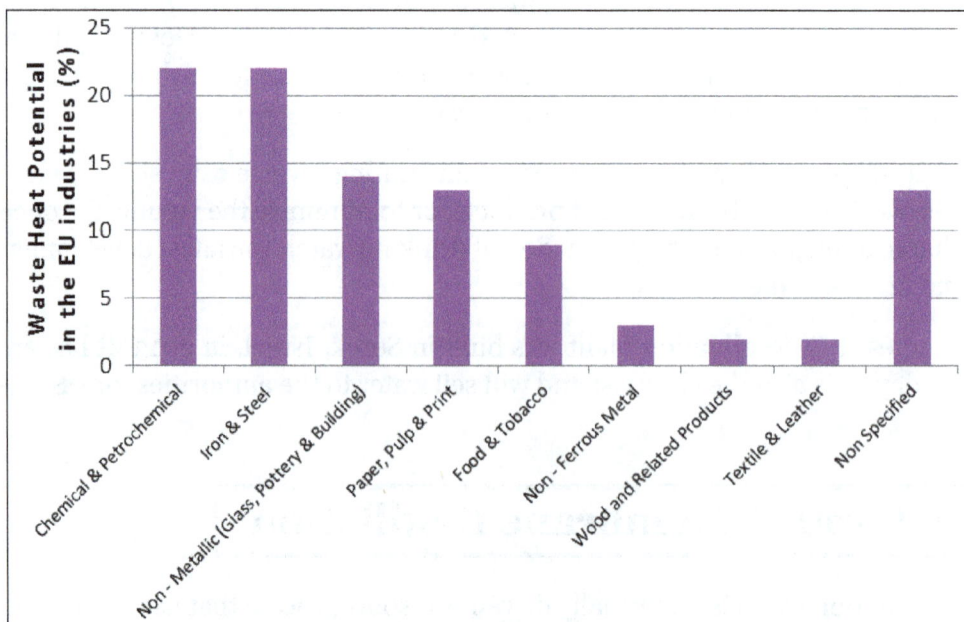

Waste heat potential per industrial sector in the EU (%), Preliminary assessment of waste heat potential in major European industries (2107).

Process Function

The driving force for MD process is given by the vapor pressure difference which is generated by a temperature difference across the membrane. As the driving force is not a pure thermal driving force, MD can be held at a much lower temperature (30-60 °C) than conventional thermal distillation. The hydrophobic nature of the membrane prevents entry to the water molecules due to surface tensions. The latter doesn't apply for the water vapors though, which create a pressure difference and travel through the membrane pore system, condensating on the opposite cool side of the membrane. The process removes ca. 85% water from the feed solution and can be summarized in three steps: (1) formation of a vapor gap at the hot feed solution–membrane interface; (2) transport of the vapor phase through the microporous system; (3) condensation of the vapor at the cold side membrane–permeate solution interface.

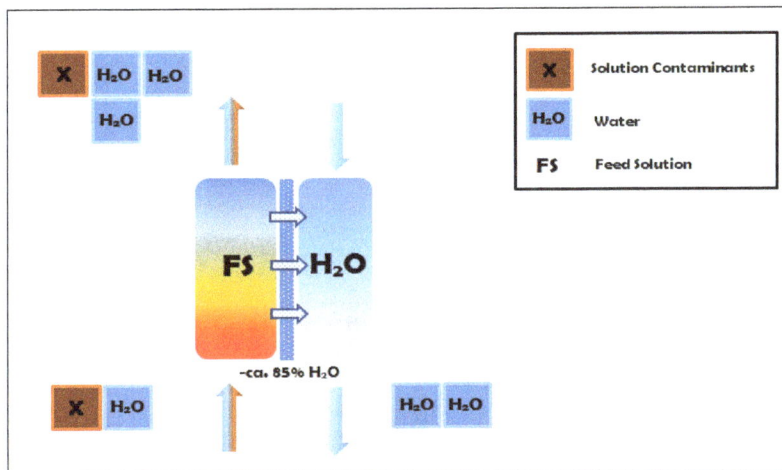

Simplified schematic of the MD process.

The way the vapor pressure difference is created across the membrane is determined by the MD module configuration. In the most commonly used configuration, direct contact membrane distillation (DCMD), the permeate-side consists of a condensation liquid (often clean water) that is in direct contact with the membrane. Alternatively, the evaporated solvent can be collected on a condensation surface that can be separated from the membrane via an air gap (AGMD) or a vacuum (VMD), or can be discharged via a cold, inert sweep gas (SGMD).

The selection of the membrane is the most crucial factor in MD separation performance. There are two common types of membrane configurations:

- Hollow fiber membrane mainly prepared from polypropylene (PP), polyvinylidenefluoride (PVDF) and PVDF - Polytetrafluoroethylene (PTFE), composite material.

- Flat sheet membrane mainly prepared from PP, PTFE, and PVDF.

PTFE has the highest hydrophobicity, good chemical and thermal stability and oxidation resistance, but it has the highest conductivity which will cause greater heat transfer through PTFE membranes (thus reducing the temperature difference and the vapor transfer). PVDF has good hydrophobicity, thermal resistance and mechanical strength and can be easily prepared into membranes with versatile pore structures. PP exhibits good thermal and chemical resistance.

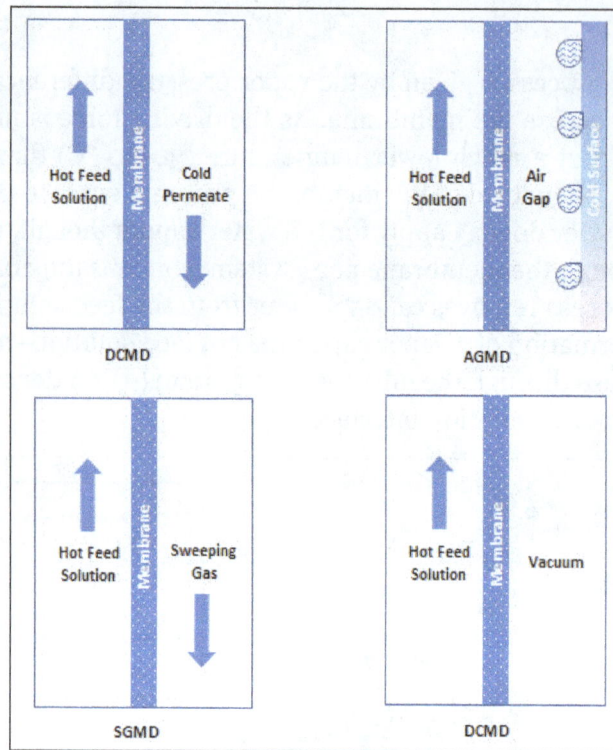

MD configurations.

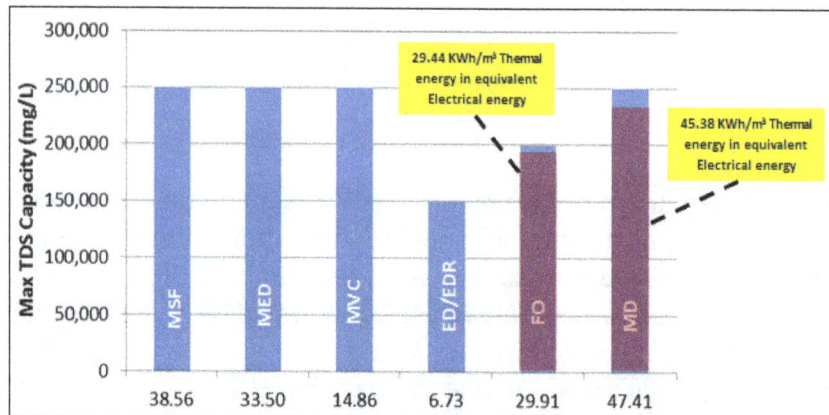

Specific Energy Consumptions (SECs) of Brine Treatment technologies in KWh/m3 versus their Max TDS Capacity in mg/L (ppm). In series we have Multi Stage Flash (MSF), Multiple Effect Distillation (MED), Mechanical Vapor Compression (MVC), Electrodialysis/Electrodialysis Reversal (ED/EDR), Forward Osmosis (FO), Membrane Distillation (MD). FO and MD can make use of waste heat for up to 90% of their Thermal Energy Demand.

Advantages and Disadvantages

Advantages:

- Low energy requirements.

- Isn't affected much by Concentration Polarization.

- 100% theoretical rejection of non-volatile components, no limit on feed concentration.

The advantages of MD, in comparison with conventional separation methods are mainly the lower pressure and the low temperature requirements (30-60°C) which lead to lower energy costs and less taxing mechanical properties for the modules. Contrary to distillation and RO the feed solution can be separated at a temperature below its boiling point (at atmospheric pressure). With the low grade heat requirements the industrial waste heat can be used, as well as renewable energy sources such as solar, wind and geothermal.

Also in comparison with RO, MD is less susceptible to flux limitations caused by concentration polarization. Very low feed temperatures can produce reasonably high rates of product water and may be more practical considering the nature of some water impurities (e.g. scaling issues at high temperature). Theoretically, MD offers 100% retention for non-volatile dissolved substances, whereby there is no limit on the supply concentration.

Disadvantages:

- Relatively high energy consumption (although the energy source is low grade temperature).

- Relatively high module cost.

- Low flux in comparison to other pressure driven membranes.

- Surfactants or amphiphilic contaminants may cause wetting of the membrane (saline feed leaks through the membrane, contaminating the permeate).

The main factors that still hinder the industrial application MD are the relatively low permeate flux in comparison with pressure-based membrane processes, flux reductions caused by concentration polarization, fouling and pore wetting of the membrane, the high cost of MD modules and the high thermal energy consumption.

Process Industry Applications

- Brine Concentration

- Cooling Towers Blowdown Treatment

- Removal of volatile components (e.g. Ammonia)

- Water purification in the pharmaceutical, chemical and textile industries

- Food & Beverages

- Resource concentration

Desalination

Desalination is a process that takes away mineral components from saline water. More generally, desalination refers to the removal of salts and minerals from a target substance, as in soil desalination, which is an issue for agriculture.

Saltwater is desalinated to produce water suitable for human consumption or irrigation. One by-product of desalination is salt. Desalination is used on many seagoing ships and submarines. Most of the modern interest in desalination is focused on cost-effective provision of fresh water for human use. Along with recycled wastewater, it is one of the few rainfall-independent water sources.

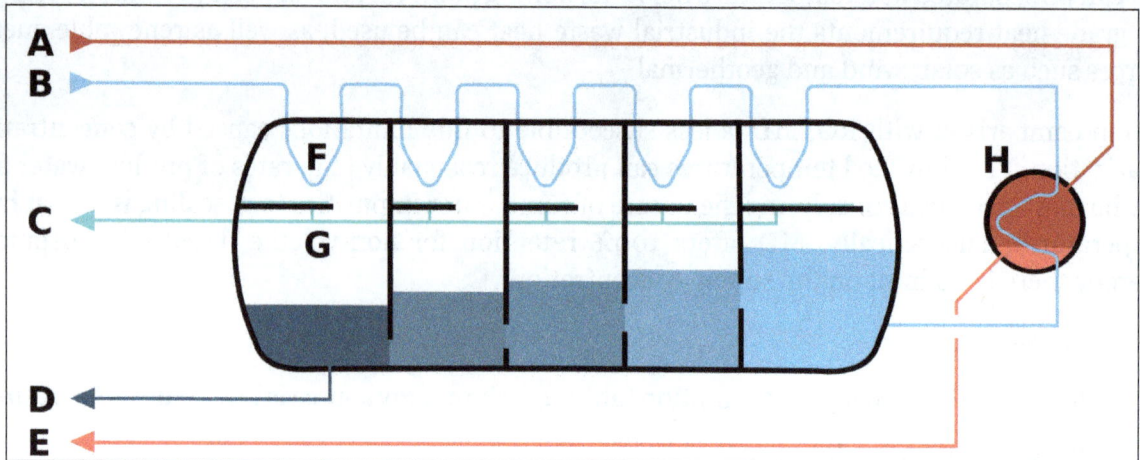

Schematic of a multistage flash desalinator
A – steam in B – seawater in C – potable water out D – brine out (waste) E – condensate out F – heat exchange G – condensation collection (desalinated water) H – brine heater.

Due to its energy consumption, desalinating sea water is generally more costly than fresh water from rivers or groundwater, water recycling and water conservation. However, these alternatives are not always available and depletion of reserves is a critical problem worldwide. Desalination processes are usually driven by either thermal (in the case of distillation) or electrical (e.g., photovoltaic or wind power) as the primary energy types.

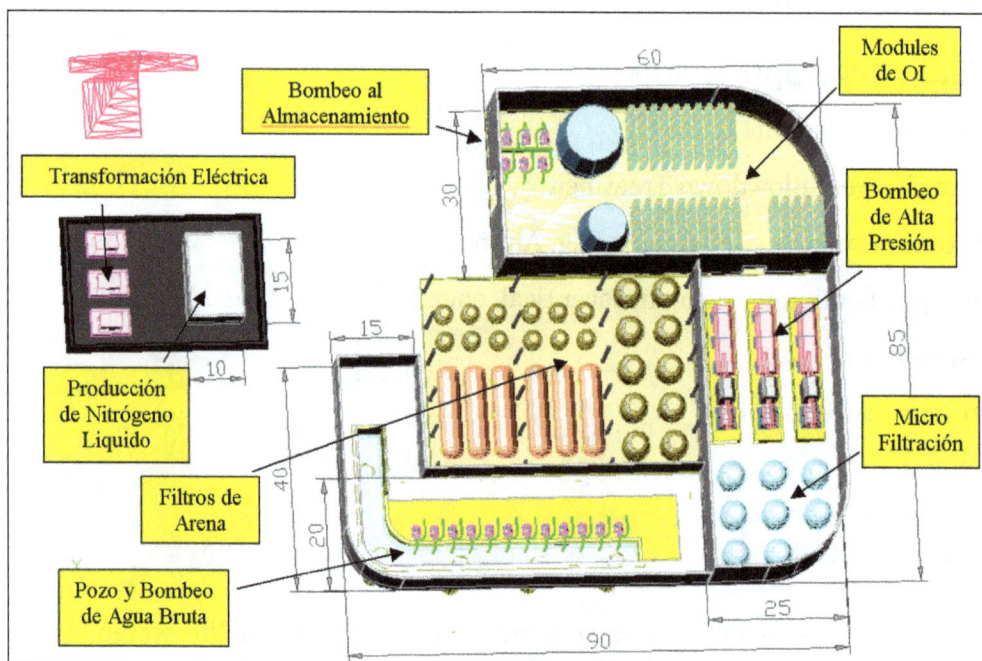

Plan of a typical reverse osmosis desalination plant.

Currently, approximately 1% of the world's population is dependent on desalinated water to meet daily needs, but the UN expects that 14% of the world's population will encounter water scarcity by 2025. Desalination is particularly relevant in dry countries such as Australia, which traditionally have relied on collecting rainfall behind dams for water.

According to the International Desalination Association, in June 2015, there were 18,426 desalination plants operated worldwide, producing 86.8 million cubic meters per day, and providing water for 300 million people. This number increased from 78.4 million cubic meters in 2013, a 10.7% increase in 2 years. The single largest desalination project is Ras Al-Khair in Saudi Arabia, which produced 1,025,000 cubic meters per day in 2014. Kuwait produces a higher proportion of its water than any other country, totaling 100% of its water use.

The pressure vessel acts as a countercurrent heat exchanger. A vacuum pump lowers the pressure in the vessel to facilitate the evaporation of the heated sea water (brine) which enters the vessel from the right side (darker shades indicate lower temperature). The steam condensates on the pipes on top of the vessel in which the fresh sea water moves from the left to the right.

Methods: There are several methods. Each has advantages and disadvantages but all are useful.

Reverse osmosis desalination plant in Barcelona, Spain.

The traditional process of desalination is distillation, i.e. boiling and re-condensation of seawater to leave salt and impurities behind.

Solar Distillation

Solar distillation mimics the natural water cycle, in which the sun heats the sea water enough for evaporation to occur. After evaporation, the water vapor is condensed onto a cool surface. There are two types of solar desalination. The former one is using photovoltaic cells which converts solar energy to electrical energy to power the desalination process. The latter one utilises the solar energy in the heat form itself and is known as solar thermal powered desalination.

Vacuum Distillation

In vacuum distillation atmospheric pressure is reduced, thus lowering the temperature required to evaporate the water. Liquids boil when the vapor pressure equals the ambient pressure and vapor pressure increases with temperature. Effectively, liquids boil at a lower temperature, when the ambient atmospheric pressure is less than usual atmospheric pressure. Thus, because of the reduced pressure, low-temperature "waste" heat from electrical power generation or industrial processes can be employed.

Multi-stage Flash Distillation

Water is evaporated and separated from sea water through multi-stage flash distillation, which is a series of flash evaporations. Each subsequent flash process utilizes energy released from the condensation of the water vapor from the previous step.

Multiple-effect Distillation

Multiple-effect distillation (MED) works through a series of steps called "effects". Incoming water is sprayed onto pipes which are then heated to generate steam. The steam is then used to heat the next batch of incoming sea water. To increase efficiency, the steam used to heat the sea water can be taken from nearby power plants. Although this method is the most thermodynamically efficient among methods powered by heat, a few limitations exist such as a max temperature and max number of effects.

Vapor-compression Distillation

Vapor-compression evaporation involves using either a mechanical compressor or a jet stream to compress the vapor present above the liquid. The compressed vapor is then used to provide the heat needed for the evaporation of the rest of the sea water. Since this system only requires power, it is more cost effective if kept at a small scale.

Freeze-thaw

Freeze-thaw desalination uses freezing to remove fresh water from salt water. Salt water is sprayed during freezing conditions into a pad where an ice-pile builds up. When seasonal conditions warm, naturally desalinated melt water is recovered. This technique relies on extended periods of natural sub-freezing conditions.

A different freeze-thaw method, not weather dependent and invented by Alexander Zarchin, freezes seawater in a vacuum. Under vacuum conditions the ice, desalinated, is melted and diverted for collection and the salt is collected.

Electrodialysis Membrane

Electrodialysis utilizes electric potential to move the salts through pairs of charged membranes, which trap salt in alternating channels.

Membrane Distillation

Membrane distillation uses a temperature difference across a membrane to evaporate vapor from a brine solution and condense pure condensate on the colder side.

Wave-powered Desalination

CETO is a wave power technology that desalinates seawater using submerged buoys. Wave-powered desalination plants began operating on Garden Island in Western Australia in 2013 and in Perth in 2015.

Energy Consumption

Energy consumption of seawater desalination has reached as low as 3 kWh/m³, including pre-filtering and ancillaries, similar to the energy consumption of other fresh water supplies transported over large distances, but much higher than local fresh water supplies that use 0.2 kWh/m³ or less.

A minimum energy consumption for seawater desalination of around 1 kWh/m³ has been determined, excluding prefiltering and intake/outfall pumping. Under 2 kWh/m³ has been achieved with reverse osmosis membrane technology, leaving limited scope for further energy reductions.

Supplying all US domestic water by desalination would increase domestic energy consumption by around 10%, about the amount of energy used by domestic refrigerators. Domestic consumption is a relatively small fraction of the total water usage.

Energy consumption of seawater desalination methods.

Desalination Method >>	Multi-stage Flash MSF	Multi-Effect Distillation MED	Mechanical Vapor Compression MVC	Reverse Osmosis RO
Electrical energy (kWh/m³)	4–6	1.5–2.5	7–12	3–5.5
Thermal energy (kWh/m³)	50–110	60–110	None	None
Electrical equivalent of thermal energy (kWh/m³)	9.5–19.5	5–8.5	None	None
Total equivalent electrical energy (kWh/m³)	13.5–25.5	6.5–11	7–12	3–5.5

"Electrical equivalent" refers to the amount of electrical energy that could be generated using a given quantity of thermal energy and appropriate turbine generator. These calculations do not include the energy required to construct or refurbish items consumed in the process.

Cogeneration

Cogeneration is generating excess heat and electricity generation from a single process. Cogeneration can provide usable heat for desalination in an integrated, or "dual-purpose", facility where a power plant provides the energy for desalination. Alternatively, the facility's energy production may be dedicated to the production of potable water (a stand-alone facility), or excess energy may be produced and incorporated into the energy grid. Cogeneration takes various forms, and theoretically any form of energy production could be used. However, the majority of current and planned cogeneration desalination plants use either fossil fuels or nuclear power as their source of energy. Most plants are located in the Middle East or North Africa, which use their petroleum resources to offset limited water resources. The advantage of dual-purpose facilities is they can be more efficient in energy consumption, thus making desalination more viable.

The Shevchenko BN-350, a former nuclear-heated desalination unit in Kazakhstan.

The current trend in dual-purpose facilities is hybrid configurations, in which the permeate from reverse osmosis desalination is mixed with distillate from thermal desalination. Basically, two or more desalination processes are combined along with power production. Such facilities have been implemented in Saudi Arabia at Jeddah and Yanbu.

A typical Supercarrier in the US military is capable of using nuclear power to desalinate 1,500,000 L of water per day.

Environmental

Factors that determine the costs for desalination include capacity and type of facility, location, feed water, labor, energy, financing and concentrate disposal.

Intake

In the United States, cooling water intake structures are regulated by the Environmental Protection Agency (EPA). These structures can have the same impacts to the environment as desalination facility intakes. According to EPA, water intake structures cause adverse environmental impact by sucking fish and shellfish or their eggs into an industrial system. There, the organisms may be killed or injured by heat, physical stress, or chemicals. Larger organisms may be killed or injured when they become trapped against screens at the front of an intake structure. Alternative intake types that mitigate these impacts include beach wells, but they require more energy and higher costs.

The Kwinana Desalination Plant opened in Perth in 2007. Water there and at Queensland's Gold Coast Desalination Plant and Sydney's Kurnell Desalination Plant is withdrawn at 0.1 m/s (0.33 ft/s), which is slow enough to let fish escape. The plant provides nearly 140,000 m³ (4,900,000 cu ft) of clean water per day.

Outflow

Desalination processes produce large quantities of brine, possibly at above ambient temperature, and contain residues of pretreatment and cleaning chemicals, their reaction byproducts and heavy metals due to corrosion. Chemical pretreatment and cleaning are a necessity in most desalination plants, which typically includes prevention of biofouling, scaling, foaming and corrosion in thermal plants, and of biofouling, suspended solids and scale deposits in membrane plants.

To limit the environmental impact of returning the brine to the ocean, it can be diluted with another stream of water entering the ocean, such as the outfall of a wastewater treatment or power plant. With medium to large power plant and desalination plants, the power plant's cooling water flow is likely to be several times larger than that of the desalination plant, reducing the salinity of the combination. Another method to dilute the brine is to mix it via a diffuser in a mixing zone. For example, once a pipeline containing the brine reaches the sea floor, it can split into many branches, each releasing brine gradually through small holes along its length. Mixing can be combined with power plant or wastewater plant dilution.

Brine is denser than seawater and therefore sinks to the ocean bottom and can damage the ecosystem. Careful reintroduction can minimize this problem.

Alternatives to Desalination

Increased water conservation and efficiency remain the most cost-effective approaches in areas with a large potential to improve the efficiency of water use practices. Wastewater reclamation provides multiple benefits over desalination. Urban runoff and storm water capture also provide benefits in treating, restoring and recharging groundwater.

A proposed alternative to desalination in the American Southwest is the commercial importation of bulk water from water-rich areas either by oil tankers converted to water carriers, or pipelines. The idea is politically unpopular in Canada, where governments imposed trade barriers to bulk water exports as a result of a North American Free Trade Agreement (NAFTA) claim.

Public Health Concerns

Desalination removes iodine from water and could increase the risk of iodine deficiency disorders. Israeli researchers claimed a possible link between seawater desalination and iodine deficiency, finding deficits among euthyroid adults exposed to iodine-poor water concurrently with an increasing proportion of their area's drinking water from seawater reverse osmosis (SWRO). They later found probable iodine deficiency disorders in a population reliant on desalinated seawater. A possible link of heavy desalinated water use and national iodine deficiency was suggested by Israeli researchers. They found a high burden of iodine deficiency in the general population of Israel: 62% of school-age children and 85% of pregnant women fall below the WHO's adequacy range. They also pointed out the national reliance on iodine-depleted desalinated water, the absence of a universal salt iodization program and reports of increased use of thyroid medication in Israel as a possible reasons that the population's iodine intake is low. In the year that the survey was conducted, the amount of water produced from the desalination plants constitutes about 50% of the quantity of fresh water supplied for all needs and about 80% of the water supplied for domestic and industrial needs in Israel.

Other Issues

Due to the nature of the process, there is a need to place the plants on approximately 25 acres of land on or near the shoreline. In the case a plant is built inland, pipes have to be laid into the ground to allow for easy intake and outtake. However, once the pipes are laid into the ground, they have a possibility of leaking into and contaminating nearby aquifers. Aside from environmental risks, the noise generated by certain types of desalination plants can be loud.

Public Opinion

Despite the issues associated with desalination processes, public support for its development can be very high. One survey of a Southern California community saw 71.9% of all respondents being in support of desalination plant development in their community. In many cases, high freshwater scarcity corresponds to higher public support for desalination development whereas areas with low water scarcity tend to have less public support for its development.

Experimental Techniques

Other desalination techniques include:

Waste Heat

Thermally-driven desalination technologies are frequently suggested for use with low-temperature waste heat sources, as the low temperatures are not useful for many industrial processes, but ideal for the lower temperatures found in desalination. In fact, such pairing with waste heat can even improve electrical process: Diesel generators commonly provide electricity in remote areas. About 40–50% of the energy output is low-grade heat that leaves the engine via the exhaust. Connecting a thermal desalination technology such as membrane distillation system to the diesel engine exhaust repurposes this low-grade heat for desalination. The system actively cools the diesel generator, improving its efficiency and increasing its electricity output. This results in an energy-neutral desalination solution. An example plant was commissioned by Dutch company Aquaver in March 2014 for Gulhi, Maldives.

Low-temperature Thermal

Originally stemming from ocean thermal energy conversion research, low-temperature thermal desalination (LTTD) takes advantage of water boiling at low pressure, even at ambient temperature. The system uses pumps to create a low-pressure, low-temperature environment in which water boils at a temperature gradient of 8–10 °C (46–50 °F) between two volumes of water. Cool ocean water is supplied from depths of up to 600 m (2,000 ft). This water is pumped through coils to condense the water vapor. The resulting condensate is purified water. LTTD may take advantage of the temperature gradient available at power plants, where large quantities of warm wastewater are discharged from the plant, reducing the energy input needed to create a temperature gradient.

Experiments were conducted in the US and Japan to test the approach. In Japan, a spray-flash evaporation system was tested by Saga University. In Hawaii, the National Energy Laboratory tested an open-cycle OTEC plant with fresh water and power production using a temperature difference of 20 C° between surface water and water at a depth of around 500 m (1,600 ft). LTTD was studied by India's National Institute of Ocean Technology (NIOT) in 2004. Their first LTTD plant opened in 2005 at Kavaratti in the Lakshadweep islands. The plant's capacity is 100,000 L (22,000 imp gal; 26,000 US gal)/day, at a capital cost of INR 50 million (€922,000). The plant uses deep water at a temperature of 10 to 12 °C (50 to 54 °F). In 2007, NIOT opened an experimental, floating LTTD plant off the coast of Chennai, with a capacity of 1,000,000 L (220,000 imp gal; 260,000 US gal)/day. A smaller plant was established in 2009 at the North Chennai Thermal Power Station to prove the LTTD application where power plant cooling water is available.

Thermoionic Process

In October 2009, Saltworks Technologies announced a process that uses solar or other thermal heat to drive an ionic current that removes all sodium and chlorine ions from the water using ion-exchange membranes.

Evaporation and Condensation for Crops

The Seawater greenhouse uses natural evaporation and condensation processes inside a greenhouse powered by solar energy to grow crops in arid coastal land.

Other Approaches

Adsorption-based desalination (AD) relies on the moisture absorption properties of certain materials such as Silica Gel.

Forward Osmosis

One process was commercialized by Modern Water PLC using forward osmosis, with a number of plants reported to be in operation.

Hydrogel based Desalination

The idea of the method is in the fact that when the hydrogel is put into contact with aqueous salt solution, it swells absorbing a solution with the ion composition different from the original one. This solution can be easily squeezed out from the gel by means of sieve or microfiltration membrane. The compression of the gel in closed system lead to change in salt concentration, whereas the compression in open system, while the gel is exchanging ions with bulk, lead to the change in the number of ions. The consequence of the compression and swelling in open and closed system conditions mimics the reverse Carnot Cycle of refrigerator machine. The only difference is that instead of heat this cycle transfers salt ions from the bulk of low salinity to a bulk of high salinity. Similarly to the Carnot cycle this cycle is fully reversible, so can in principle work with an ideal thermodynamic efficiency. Because the method is free from the use of osmotic membranes it can compete with reverse osmosis method. In addition, unlike the reverse osmosis, the approach is not sensitive to the quality of feed water and its seasonal changes, and allows the production of water of any desired concentration.

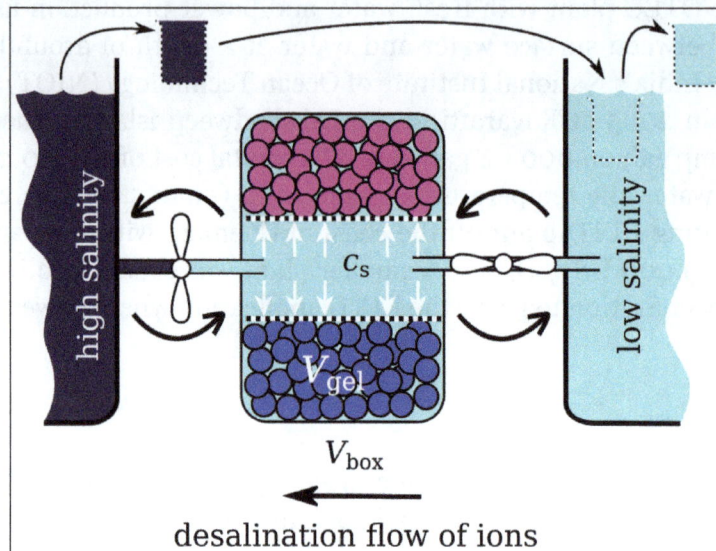

Scheme of the desalination machine: the desalination box of volume V_{box} contains a gel of volume V_{gel} which is separated by a sieve from the outer solution volume $V_{out} = V_{box} - V_{gel}$. The box is connected to two big tanks with high and low salinity by two taps which can be opened and closed as desired. The chain of buckets expresses the fresh water consumption followed by refilling the low-salinity reservoir by salt water.

Small-scale Solar

The United States, France and the United Arab Emirates are working to develop practical solar desalination. AquaDania's WaterStillar has been installed at Dahab, Egypt, and in Playa del Carmen, Mexico. In this approach, a solar thermal collector measuring two square metres can distill from 40 to 60 litres per day from any local water source – five times more than conventional stills. It eliminates the need for plastic PET bottles or energy-consuming water transport. In Central California, a startup company WaterFX is developing a solar-powered method of desalination that can enable the use of local water, including runoff water that can be treated and used again. Salty groundwater in the region would be treated to become freshwater, and in areas near the ocean, seawater could be treated.

Passarell

The Passarell process uses reduced atmospheric pressure rather than heat to drive evaporative desalination. The pure water vapor generated by distillation is then compressed and condensed using an advanced compressor. The compression process improves distillation efficiency by creating the reduced pressure in the evaporation chamber. The compressor centrifuges the pure water vapor after it is drawn through a demister (removing residual impurities) causing it to compress against tubes in the collection chamber. The compression of the vapor increases its temperature. The heat is transferred to the input water falling in the tubes, vaporizing the water in the tubes. Water vapor condenses on the outside of the tubes as product water. By combining several physical processes, Passarell enables most of the system's energy to be recycled through its evaporation, demisting, vapor compression, condensation, and water movement processes.

Geothermal

Geothermal energy can drive desalination. In most locations, geothermal desalination beats using scarce groundwater or surface water, environmentally and economically.

Nanotechnology

Nanotube membranes of higher permeability than current generation of membranes may lead to eventual reduction in the footprint of RO desalination plants. It has also been suggested that the use of such membranes will lead to reduction in the energy needed for desalination.

Hermetic, sulphonated nano-composite membranes have shown to be capable of removing a various contaminants to the parts per billion level. s, have little or no susceptibility to high salt concentration levels.

Biomimesis

Biomimetic membranes are another approach.

Electrochemical

In 2008, Siemens Water Technologies announced technology that applied electric fields to desalinate one cubic meter of water while using only a purported 1.5 kWh of energy. If accurate, this process would consume one-half the energy of other processes. As of 2012 a demonstration plant was operating in Singapore. Researchers at the University of Texas at Austin and the University of Marburg are developing more efficient methods of electrochemically mediated seawater desalination.

Electrokinetic Shocks

A process employing electrokinetic shocks waves can be used to accomplish membraneless desalination at ambient temperature and pressure. In this process, anions and cations in salt water are exchanged for carbonate anions and calcium cations, respectively using electrokinetic shockwaves. Calcium and carbonate ions react to form calcium carbonate, which precipitates, leaving fresh water. The theoretical energy efficiency of this method is on par with electrodialysis and reverse osmosis.

Temperature Swing Solvent Extraction

Temperature Swing Solvent Extraction (TSSE) uses a solvent instead of a membrane or high temperatures.

Solvent extraction is a common technique in chemical engineering. It can be activated by low-grade heat (less than 70 °C (158 °F), which may not require active heating. In a study, TSSE removed up to 98.4 percent of the salt in brine. A solvent whose solubility varies with temperature is added to saltwater. At room temperature the solvent draws water molecules away from the salt. The water-laden solvent is then heated, causing the solvent to release the now salt-free water.

It can desalinate extremely salty brine up to seven times as salty as the ocean. For comparison, the current methods can only handle brine twice as salty.

Facilities

Estimates vary widely between 15,000–20,000 desalination plants producing more than 20,000 m³/day. Micro desalination plants operate near almost every natural gas or fracking facility found in the United States.

In Nature

Evaporation of water over the oceans in the water cycle is a natural desalination process. The formation of sea ice produces ice with little salt, much lower than in seawater.

Mangrove leaf with salt crystals.

Seabirds distill seawater using countercurrent exchange in a gland with a rete mirabile. The gland secretes highly concentrated brine stored near the nostrils above the beak. The bird then "sneezes" the brine out. As freshwater is not usually available in their environments, some seabirds, such as pelicans, petrels, albatrosses, gulls and terns, possess this gland, which allows them to drink the salty water from their environments while they are far from land.

Mangrove trees grow in seawater; they secrete salt by trapping it in parts of the root, which are then eaten by animals (usually crabs). Additional salt is removed by storing it in leaves that fall off. Some types of mangroves have glands on their leaves, which work in a similar way to the seabird desalination gland. Salt is extracted to the leaf exterior as small crystals, which then fall off the leaf.

Willow trees and reeds absorb salt and other contaminants, effectively desalinating the water. This is used in artificial constructed wetlands, for treating sewage.

In Situ Chemical Oxidation

In situ chemical oxidation (ISCO), a form of advanced oxidation processes and advanced oxidation technology, is an environmental remediation technique used for soil and/or groundwater remediation to reduce the concentrations of targeted environmental contaminants to acceptable levels. ISCO is accomplished by injecting or otherwise introducing strong chemical oxidizers directly into the contaminated medium (soil or groundwater) to destroy chemical contaminants in place. It can be used to remediate a variety of organic compounds, including some that are resistant to natural degradation.

Chemical oxidation is one half of a redox reaction, which results in the loss of electrons. One of the reactants in the reaction becomes oxidized, or loses electrons, while the other reactant becomes reduced, or gains electrons. In ISCO, oxidizing compounds, compounds that give electrons away to other compounds in a reaction, are used to change the contaminants into harmless compounds. The *in situ* in ISCO is just Latin for "in place", signifying that ISCO is a chemical oxidation reaction that occurs at the site of the contamination.

The remediation of certain organic substances such as chlorinated solvents (trichloroethene and tetrachloroethene), and gasoline-related compounds (benzene, toluene, ethylbenzene, MTBE, and xylenes) by ISCO is possible. Some other contaminants can be made less toxic through chemical oxidation.

A wide range of ground water contaminants react either moderately or highly with the ISCO method, and ISCO can also be used in a variety of different situations (e.g. unsaturated vs saturated ground, above ground or underground, etc.), so it is a popular method to use.

Agents of Oxidization

Common oxidants used in this process are permanganate (both sodium permanganate and potassium permanganate), Fenton's Reagent, persulfate, and ozone. Other oxidants can be used, but these four are the most commonly used.

Permanganate

Permanganate is used in groundwater remediation in the form of potassium permanganate ($KMnO_4$) and sodium permanganate ($NaMnO_4$). Both compounds have the same oxidizing capabilities and limitations and react similarly to contaminants. The biggest difference between the two chemicals is that potassium permanganate is less soluble than sodium permanganate.

Potassium permanganate is a crystalline solid that is typically dissolved in water before application to the contaminated site. Unfortunately, the solubility of potassium permanganate is dependent on temperature. Because the temperature in the aquifer is usually less than the temperature in the area that the solution is mixed, the potassium permanganate becomes a solid material again. This solid material then does not react with the contaminants. Over time, the permanganate will become soluble again, but the process takes a long time. This compound has been shown to oxidize many different contaminants but is notable for oxidizing chlorinated solvents such as perchloroethylene (PCE), trichloroethylene (TCE), and vinyl chloride (VC). However, potassium permanganate is unable to efficiently oxidize diesel, gasoline, or BTEX.

Sodium permanganate is more expensive than potassium permanganate, but because sodium permanganate is more soluble than potassium permanganate, it can be applied to the site of contamination at a much higher concentration. This shortens the time required for the contaminant to be oxidized. Sodium permanganate is also useful in that it can be used in places where the potassium ion cannot be used. Another advantage that sodium permanganate has over potassium permanganate is that sodium permanganate, due to its high solubility, can be transported above ground as a liquid, decreasing the risk of exposure to granules or skin contact with the substance.

The primary redox reactions for permanganate are given by the equations:

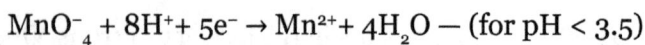

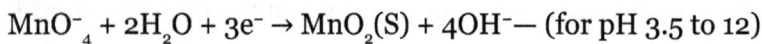

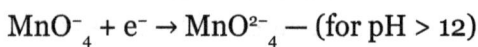

$$MnO^-_4 + 8H^+ + 5e^- \rightarrow Mn^{2+} + 4H_2O - \text{(for pH < 3.5)}$$

$$MnO^-_4 + 2H_2O + 3e^- \rightarrow MnO_2(S) + 4OH^- - \text{(for pH 3.5 to 12)}$$

$$MnO^-_4 + e^- \rightarrow MnO^{2-}_4 - \text{(for pH > 12)}$$

The typical reaction that occurs under common environmental conditions is equation 2. This reaction forms a solid product, MnO_2.

The advantage of using permanganate in ISCO is that it reacts comparatively slowly in the subsurface which allows the compound to move further into the contaminated space and oxidize more contaminants. Permanganate can also help with the cleanup of materials that are not very permeable. In addition, because both sodium permanganate and potassium permanganate solutions have a density greater than water's density, permanganate can travel through the contaminated area through density-driven diffusion.

The use of permanganate creates the byproduct MnO_2, which is naturally present in the soil and is therefore a safe byproduct. Unfortunately, several studies have shown that this byproduct seems to cement sand particles together forming rock-like material that has very low permeability. As the rock-like materials build up, it blocks the permanganate from getting to the rest of the contaminant and lowers the efficiency of the permanganate. This can be prevented by extracting the MnO_2 from the contaminated area.

Fenton's Reagent

Fenton's reagent is basically a mixture of ferrous iron salts as a catalyst and hydrogen peroxide. A similar sort of reaction can be made by mixing hydrogen peroxide with [ferric] iron (Iron III). When the peroxide is catalyzed by soluble iron it forms hydroxyl radicals($\cdot$OH) that oxidize contaminants such as chlorinated solvents, fuel oils, and BTEX. Traditional Fenton's reagent usually requires a significant pH reduction of the soils and groundwater in the treatment zone to allow for the introduction and distribution of aqueous iron as iron will oxidize and precipitate at a pH greater than 3.5. Unfortunately, the contaminated groundwater that needs to be treated has a pH level that is at or near neutral. Due to this, there are controversies on whether ISCO using Fenton's reagent is really a Fenton reaction. Instead, scientists call these reactions Fenton-like. However, some ISCO vendors successfully apply pH neutral Fenton's reagent by chelating the iron which keeps the iron in solution and mitigates the need for acidifying the treatment zone. The Fenton chemistry is complex and has many steps, including the following:

$$Fe^{2+} + H_2O_2 \rightarrow Fe^{3+} + OH\cdot + OH^-$$

$$Fe^{3+} + H_2O_2 \rightarrow Fe^{2+} + OOH\cdot + H^+$$

$$HO\cdot + H_2O_2 \rightarrow Fe(III) + HO_2^{\cdot} + H^+$$

$$HO\cdot + Fe(II) \rightarrow Fe(III) + OH^-$$

$$Fe(III) + HO_2^{\cdot} \rightarrow Fe(II) + O_2H^+$$

$$Fe(II) + HO_2^{\cdot} + H^+ \rightarrow Fe(III) + H_2O_2$$

$$HO_2^{\cdot} + HO_2^{\cdot} \rightarrow H_2O_2 + O_2$$

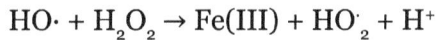

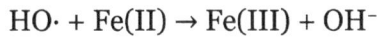

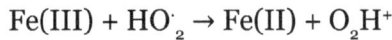

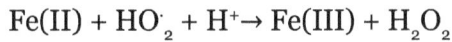

These reactions do not occur step by step but simultaneously.

When applied to In Situ Chemical Oxidation, the collective reaction results in the degradation of contaminants in the presence of Fe^{2+} as a catalyst. The overall end result of the process can be described by the following reaction:

$$H_2O_2 + contaminant \rightarrow H_2O + CO_2 + O_2$$

Advantages of this method include that the hydroxyl radicals are very strong oxidants and react very rapidly with contaminants and impurities in the ground water. Moreover, the chemicals needed for this process are inexpensive and abundant.

Traditional Fenton's reagent applications can be very exothermic when significant iron, manganese or contaminant (i.e. NAPL concentrations) are present in an injection zone. Over the course of the reaction, the groundwater heats up and, in some cases, reagent and vapors can surface out of the soil. Stabilizing the peroxide can significantly increase the residence time and distribution of the reagent while reducing the potential for excessive temperatures by effectively isolating the peroxide from naturally occurring divalent transition metals in the treatment zone. However, NAPL contaminant concentrations can still result in rapid oxidation reactions with an associated temperature increase and more potential for surfacing even with reagent stabilization. The hydroxyl radicals can be scavenged by carbonate, bicarbonate, and naturally occurring organic matter in addition to the targeted contaminant, so it important to evaluate a site's soil matrix and apply additional reagent when these soil components are present in significant abundance.

Persulfate

Persulfate is a newer oxidant used in ISCO technology. The persulfate compound that is used in groundwater remediation is in the form of peroxodisulfate or peroxydisulfate ($S_2O_8^{2-}$) but is generally called a persulfate ion by scientists in the field of environmental engineering. More specifically, sodium persulfate is used because it has the highest water solubility and its reaction with contaminants leaves least harmful side products. Although sodium persulfate by itself can degrade many environmental contaminants, the sulfate radical SO_4^- is usually derived from the persulfate because sulfate radicals can degrade a wider range of contaminants at a faster pace(about 1,000–100,000 times) than the persulfate ion. Various agents, such as heat, ultraviolet light, high pH, hydrogen peroxide, and transition metals, are used to activate persulfate ions and generate sulfate radicals.

The sulfate radical is an electrophile, a compound that is attracted to electrons and that reacts by accepting an electron pair in order to bond to a nucleophile. Therefore the performance of sulfate radicals is enhanced in an area where there are many electron donating organic compounds. The sulfate radical reacts with the organic compounds to form an organic radical cation. Examples of electron donating groups present in organic compounds are the amino (-NH_2), hydroxyl (-OH), and alkoxy (-OR) groups. Conversely, the sulfate radical does not react as much in compounds that contain electron attracting groups like nitro (-NO_2) and carbonyl (C=O) and also in the presence of substances containing chlorine atoms. Also, as the number of ether bonds increases, the reaction rates decrease.

When applied in the field, persulfate must first be activated (it must be turned into the sulfate radical) to be effective in the decontamination. The catalyst that is most commonly used is ferrous iron (Iron II). When ferrous iron and persulfate ions are mixed together, they produce ferric iron (iron III) and two types of sulfate radicals, one with a charge of –1 and the other with a charge of –2. New research has shown that Zero Valent Iron (ZVI) can also be used with persulfate with success. The persulfate and the iron are not mixed beforehand, but are injected into the area of contamination together. The persulfate and iron react underground to produce the sulfate radicals. The rate of contaminant destruction increases as the temperature of the surroundings increases.

The advantage of using persulfate is that persulfate is much more stable than either hydrogen peroxide or ozone above the surface and it does not react quickly by nature. This means fewer transportation limitations, it can be injected into the site of contamination at high concentrations, and can be transported through porous media by density driven diffusion. The disadvantage is that this is an emerging field of technology and there are only a few reports of testing it in the field and more research needs to be done with it. Additionally, each mole of persulfate creates one mole of oxidizer (sulfate radical or hydroxyl radical). These radicals have low atomic weights while the persulfate molecule has a high atomic weight (238). Therefore, the value (oxidizer produced when persulfate is activated) for expense (price of relatively heavy persulfate molecule) is low compared to some other oxidizing reagents.

Ozone

While oxygen is a very strong oxidant, it's elemental form O_2 is not very soluble in water. This poses a problem in ground water remediation, because the chemical must be able to mix with water to remove the contaminant. Fortunately, ozone (O_3) is about 12 times more soluble than O_2 and, although it is still comparably insoluble, it is a strong oxidant.

The unique part of ozone oxidation is its in-situ application. Because, unlike other oxidants used in ISCO, it is a gas, it needs to be injected into the contamination site from the bottom rather than the top. Tubes are built into the ground to transport the ozone to its starting place; the bubbles then rise to the surface. Whatever volatile substances are left over are sucked up by a vacuum pump. Because the bubbles travel more vertically than horizontally, close placement of ozone injection wells is needed for uniform distribution.

The biggest advantage in using ozone in ISCO is that ozone does not leave any residual chemical like persulfate leaves SO_4^{2-} or permanganate leaves MnO_2. The processes involved with ozonation (treating water with ozone) only leave behind O_2. Ozone can also react with many of the important environmental contaminants. In addition, because ozone is a gas, adding ozone to the bottom of the contaminant pool forces the ozone to rise up through the contaminants and react. Because of this property, ozone can also be delivered more quickly. Also, in theory, H_2O_2 co-injected with ozone will result in -OH ions, which are very strong oxidants.

However, ozone has many properties that pose problems. Ozone reacts with a variety of contaminants, but the problem is that it also reacts quickly with many other substances such as minerals, organic matter, etc. that are not the targeted substances. Again, it is not very soluble and stays in gas form in the water, which makes ozone prone to nonuniform distribution and rising up to the top of contamination site by the shortest routes rather than traveling through the entire material. In addition, ozone must be generated, and that requires a huge amount of energy.

Implementation

The primary delivery mechanism for ISCO is through perforated, hollow metal rods hammered into the ground by "direct-push" drilling methods or by injecting the oxidant into wells installed using hollow stem auger, rotary drilling methods. One advantage of injection wells is that they can be used for multiple applications of the oxidant material, while direct push injection techniques are generally quicker and less expensive. Injection wells for ozone are typically constructed of a 1–2" stainless-steel screen set in sand pack, grouted to the surface using a combination of cement and bentonite clay. Often, a field pilot study must be performed to determine injection parameters and well spacing.

Oxidants such as permanganate and Fenton's Reagent are delivered as water-based solutions. These substances are injected into the aquifer and then allowed to propagate by gravity and water current. As contaminants are encountered, the substances oxidize them and purify the water. Ozone is delivered (sparged) as a gas in either a dry air or oxygen carrier gas. Specialized equipment is required for in-situ oxidation via ozone gas injection. The ozone has to be pumped into the groundwater from the bottom of the aquifer because the ozone gas is less dense than the water. As the ozone travels through the aquifer against gravity, it reacts with contaminants along the way. However, there are some specific methods of oxidant delivery including injection probes, hydraulic fracturing, soil mixing, vertical wells, horizontal wells, and treatment walls.

Injection Probes

Injection probes are used in areas where there is very low permeability. A small diameter probe (2 to 4 cm in diameter) is rotated or pushed into the ground while reagents are inserted into it at low pressure. The reagents travel down the core of the probe and exit out though small perforations along the sides of the probe which are located at certain intervals. The reagents travel away from the core by going into existing cracks and pores and create a "halo of reactivity" (from pg. 182 or *Principles and Practices of In Situ Chemical Oxidation Using Permanganate*). In order to optimize the amount of contaminant that is oxidized, the probes are set into the ground relatively close together, about .6-1.2 meters apart.

Hydraulic Fracturing

Hydraulic fracturing is the process of artificially creating fractures in a site that has low permeability and then filling the fractures with oxidants. First a hole is drilled into the ground, and then a forceful jet of water is used to create fractures. Coarse sand, which allows just enough permeability for oxidants to get though, is used to fill the fractures and prevent them from closing up, and after that, the oxidant is injected into the fracture.

Soil Mixing

Soil mixing can be used to deliver solid or liquid forms of oxidants to contaminated soil. For near surface to intermediate contamination zones, either standard construction equipment (i.e. bucket mixing), or specialized soil mixing tools (i.e. Lang Tool, Allu Tool, Alpine, etc.) can be used. Deep soil mixing requires specialized auger mixing equipment. In order to apply this method in-situ and in deep soil, the oxidant must be pumped to the point of mixing using a kelly bar (a piece of earth drilling equipment), or appropriate piping to the place where the soil needs to be oxidized. The soil then has to be mixed by using mixing blades.

Horizontal and Vertical Wells

Horizontal well networks are basically the use of long pipes that lead in and out of the contaminated aquifer or plume used to inject oxidants and extract the treated ground water. Vertical wells networks consist of appropriately spaced injection wells with slightly overlapping radius of influence (ROI) to ensure reagent contact within the vertical and horizontal treatment zone. Injection wells can be permanently installed or be temporarily installed (i.e. by using direct push technology). Horizontal well networks use pipes that are slightly L-shaped at the bottom to inject oxidant and extract treated groundwater horizontally. Horizontal wells are used especially when oxidants need to be delivered to thin layers of saturation.

Treatment Walls

Treatment walls are used to deliver oxidants to the end of a contaminant plume and can be used to prevent the migration of an oxidant. The walls usually consist of continuous trenches that are connected to a piping network into which oxidants can be injected into. Another version of this delivery system is the use of a disconnected series of vertical wells to inject the oxidant into the ground water. The factors that affect treatment wall application and performance are similar to the factors that effect the performance of permeable reactive barriers.

Performance Application

The effectiveness of the oxidation is contingent on the site lithology, the residence time of the oxidant, the amount of oxidant used, the presence of oxidizing materials other than the targeted contaminant, the degree of effective contact between the oxidant and the contaminant(s), and the kinetics of the oxidation reaction between the oxidant and contaminant.

The soil and groundwater are tested both before and after oxidant application to verify the effectiveness of the process. Monitoring of gases given off during oxidation can also help determine if contaminants are being destroyed. Elevated levels of CO_2 is an indicator of oxidation.

Safety and Hazards

The four main types of oxidants that are used in ISCO—Fenton's reagent, ozone, permanganate, and persulfate—are all strong oxidizing agents and pose serious hazards to the people who are working with them. For worker safety, site that are using ozone as the oxidant must test ozone levels in the air periodically because ozone has adverse respiratory effects. All oxidants must be stored properly so that they do not decompose and workers must ensure that they do not have skin contact with any of the oxidants.

Some ISCO compounds can react aggressively with organic contaminants and must be used with care on the site. Fenton's reagent in particular is highly exothermic and can cause unwanted effects on microbial life in the aquifer if it is not used carefully or stabilized.

Further challenges associated with ISCO include the generation of unwanted or toxic oxidation products. Recent evidence suggests that the oxidation of benzene results in the formation of phenol (a relatively benign compound) and a novel aldehyde side-product, the toxicology of which is unknown.

Potential Improvements

Currently ISCO is mostly applied by itself, but it may be possible to combine ISCO with other technologies such as in situ chemical reduction (ISCR) and in situ thermal desorption (ISTD). As ISCO is not efficient at treating low concentration contaminant plumes, ISCO can be used to treat the contaminant source while ISCR treats the plumes.

Traditional ISCO is limited by mass transfer of contaminants into the aqueous (groundwater) phase. Since the oxidation reaction takes place in the groundwater, contaminant destruction is restricted to only those contaminants which have partitioned into the groundwater phase. To overcome this limitation at sites which have substantial soil contamination, and/or non-aqueous phase liquid (NAPL), surfactants can be injected simultaneously with oxidants. The surfactants emulsify soil sorbed contaminants and/or NAPL enabling them to be destroyed in aqueous phase oxidative reactions; this patented technology is known as Surfactant-enhanced In Situ Chemical Oxidation (S-ISCO).

The ISCO delivery technology and reagents also could be enhanced. Currently, an oxidant is injected into the contaminated site and is distributed by the injection pressure, turbulence and advection. This method is effective with appropriate point spacing and slightly overlapping radius of influence (ROI). However, peroxide based reagents are not very stable and react with other substances soon after being injected into the sub-surface unless the peroxide is stabilized. Additionally, current persulfate activation methods often stall resulting in sub-optimal results. These problems could be fixed by creating oxidants that are more stable and specifically targeted to contaminants, so that they do not oxidize other substances. The delivery systems could also be improved so that the oxidants are sent to the correct locations.

Bioremediation

Bioremediation is a process used to treat contaminated media, including water, soil and subsurface material, by altering environmental conditions to stimulate growth of microorganisms and degrade the target pollutants. In many cases, bioremediation is less expensive and more sustainable than other remediation alternatives. Biological treatment is a similar approach used to treat wastes including wastewater, industrial waste and solid waste.

Most bioremediation processes involve oxidation-reduction reactions where either an electron acceptor (commonly oxygen) is added to stimulate oxidation of a reduced pollutant (e.g. hydrocarbons) or an electron donor (commonly an organic substrate) is added to reduce oxidized pollutants (nitrate, perchlorate, oxidized metals, chlorinated solvents, explosives and propellants). In both these approaches, additional nutrients, vitamins, minerals, and pH buffers may be added to optimize conditions for the microorganisms. In some cases, specialized microbial cultures are added (bioaugmentation) to further enhance biodegradation. Some examples of bioremediation related technologies are phytoremediation, mycoremediation, bioventing, bioleaching, landfarming, bioreactor, composting, bioaugmentation, rhizofiltration, and biostimulation.

Chemistry

Most bioremediation processes involve oxidation-reduction (Redox) reactions where a chemical species donates an electron (electron donor) to a different species that accepts the electron (electron acceptor). During this process, the electron donor is said to be oxidized while the electron acceptor is reduced. Common electron acceptors in bioremediation processes include oxygen, nitrate, manganese (III and IV), iron (III), sulfate, carbon dioxide and some pollutants (chlorinated solvents, explosives, oxidized metals, and radionuclides). Electron donors include sugars, fats, alcohols, natural organic material, fuel hydrocarbons and a variety of reduced organic pollutants. The redox potential for common biotransformation reactions is shown in the table.

Process	Reaction	Redox potential (E_h in mV)
Aerobic	$O_2 + 4e^- + 4H^+ \rightarrow 2H_2O$	600 ~ 400
Anaerobic		
Denitrification	$2NO_3^- + 10e^- + 12H^+ \rightarrow N_2 + 6H_2O$	500 ~ 200
Manganese IV reduction	$MnO_2 + 2e^- + 4H^+ \rightarrow Mn^{2+} + 2H_2O$	400 ~ 200
Iron III reduction	$Fe(OH)_3 + e^- + 3H^+ \rightarrow Fe^{2+} + 3H_2O$	300 ~ 100
Sulfate reduction	$SO_4^{2-} + 8e^- + 10H^+ \rightarrow H_2S + 4H_2O$	0 ~ −150
Fermentation	$2CH_2O \rightarrow CO_2 + CH_4$	−150 ~ −220

Aerobic

Aerobic bioremediation is the most common form of oxidative bioremediation process where oxygen is provided as the electron acceptor for oxidation of petroleum, polyaromatic hydrocarbons (PAHs), phenols, and other reduced pollutants. Oxygen is generally the preferred electron acceptor because of the higher energy yield and because oxygen is required for some enzyme systems to initiate the degradation process. Numerous laboratory and field studies have shown that microorganisms can degrade a wide variety of hydrocarbons, including components of gasoline, kerosene, diesel, and jet fuel. Under ideal conditions, the biodegradation rates of the low- to moderate-weight aliphatic, alicyclic, and aromatic compounds can be very high. As the molecular weight of the compound increases, so does the resistance to biodegradation.

Common approaches for providing oxygen above the water table include landfarming, composting and bioventing. During landfarming, contaminated soils, sediments, or sludges are incorporated into the soil surface and periodically turned over (tilled) using conventional agricultural equipment to aerate the mixture. Composting accelerates pollutant biodegradation by mixing the waste to be treated with a bulking agent, forming into piles, and periodically mixed to increase oxygen transfer. Bioventing is a process that increases the oxygen or air flow into the unsaturated zone of the soil which increases the rate of natural in situ degradation of the targeted hydrocarbon contaminant.

Approaches for oxygen addition below the water table include recirculating aerated water through the treatment zone, addition of pure oxygen or peroxides, and air sparging. Recirculation systems typically consist of a combination of injection wells or galleries and one or more recovery wells where the extracted groundwater is treated, oxygenated, amended with nutrients and reinjected. However, the amount of oxygen that can be provided by this method is limited by the low solubility of oxygen in water (8 to 10 mg/L for water in equilibrium with air at typical temperatures). Greater amounts of oxygen can be provided by contacting the water with pure oxygen or addition of hydrogen peroxide (H_2O_2) to the water. In some cases, slurries of solid calcium or magnesium peroxide are injected under pressure through soil borings. These solid peroxides react with water releasing H_2O_2 which then decomposes releasing oxygen. Air sparging involves the injection of air under pressure below the water table. The air injection pressure must be great enough to overcome the hydrostatic pressure of the water and resistance to air flow through the soil.

Anaerobic

Anaerobic bioremediation can be employed to treat a broad range of oxidized contaminants including chlorinated ethenes (PCE, TCE, DCE, VC), chlorinated ethanes (TCA, DCA), chloromethanes (CT, CF), chlorinated cyclic hydrocarbons, various energetics (e.g., perchlorate, RDX, TNT), and nitrate. This process involves the addition of an electron donor to: 1) deplete background electron acceptors including oxygen, nitrate, oxidized iron and manganese and sulfate; and 2) stimulate the biological and/or chemical reduction of the oxidized pollutants. Hexavalent chromium (Cr[VI]) and uranium (U[VI]) can be reduced to less mobile and/or less toxic forms (e.g., Cr[III], U[IV]). Similarly, reduction of sulfate to sulfide (sulfidogenesis) can be used to precipitate certain metals (e.g., zinc, cadmium). The choice of substrate and the method of injection depend on the contaminant type and distribution in the aquifer, hydrogeology, and remediation objectives. Substrate can be added using conventional well installations, by direct-push technology, or by excavation and backfill such as permeable reactive barriers (PRB) or biowalls. Slow-release products composed of edible oils or solid substrates tend to stay in place for an extended treatment period. Soluble substrates or soluble fermentation products of slow-release substrates can potentially migrate via advection and diffusion, providing broader but shorter-lived treatment zones. The added organic substrates are first fermented to hydrogen (H_2) and volatile fatty acids (VFAs). The VFAs, including acetate, lactate, propionate and butyrate, provide carbon and energy for bacterial metabolism.

Heavy Metals

Heavy metals including cadmium, chromium, lead and uranium are elements so they cannot be biodegraded. However, bioremediation processes can potentially be used to reduce the mobility of these material in the subsurface, reducing the potential for human and environmental exposure.

The mobility of certain metals including chromium (Cr) and uranium (U) varies depending on the oxidation state of the material. Microorganisms can be used to reduce the toxicity and mobility of chromium by reducing hexavalent chromium, Cr(VI) to trivalent Cr (III). Uranium can be reduced from the more mobile U(VI) oxidation state to the less mobile U(IV) oxidation state. Microorganisms are used in this process because the reduction rate of these metals is often slow unless catalyzed by microbial interactions Research is also underway to develop methods to remove metals from water by enhancing the sorption of the metal to cell walls. This approach has been evaluated for treatment of cadmium, chromium, and lead.

Additives

In the event of biostimulation, adding nutrients that are limited to make the environment more suitable for bioremediation, nutrients such as nitrogen, phosphorus, oxygen, and carbon may be added to the system to improve effectiveness of the treatment.

Many biological processes are sensitive to pH and function most efficiently in near neutral conditions. Low pH can interfere with pH homeostasis or increase the solubility of toxic metals. Microorganisms can expend cellular energy to maintain homeostasis or cytoplasmic conditions may change in response to external changes in pH. Some anaerobes have adapted to low pH conditions through alterations in carbon and electron flow, cellular morphology, membrane structure, and protein synthesis.

Limitations

Bioremediation can be used to completely mineralize organic pollutants, to partially transform the pollutants, or alter their mobility. Heavy metals and radionuclides are elements that cannot be biodegraded, but can be bio-transformed to less mobile forms. In some cases, microbes do not fully mineralize the pollutant, potentially producing a more toxic compound. For example, under anaerobic conditions, the reductive dehalogenation of TCE may produce dichloroethylene (DCE) and vinyl chloride (VC), which are suspected or known carcinogens. However, the microorganism *Dehalococcoides* can further reduce DCE and VC to the non-toxic product ethene. Additional research is required to develop methods to ensure that the products from biodegradation are less persistent and less toxic than the original contaminant. Thus, the metabolic and chemical pathways of the microorganisms of interest must be known. In addition, knowing these pathways will help develop new technologies that can deal with sites that have uneven distributions of a mixture of contaminants.

Also, for biodegradation to occur, there must be a microbial population with the metabolic capacity to degrade the pollutant, an environment with the right growing conditions for the microbes, and the right amount of nutrients and contaminants. The biological processes used by these microbes are highly specific, therefore, many environmental factors must be taken into account and regulated as well. Thus, bioremediation processes must be specifically made in accordance to the conditions at the contaminated site. Also, because many factors are interdependent, small-scale tests are usually performed before carrying out the procedure at the contaminated site. However, it can be difficult to extrapolate the results from the small-scale test studies into big field operations. In many cases, bioremediation takes more time than other alternatives such as land filling and incineration.

Genetic Engineering

The use of genetic engineering to create organisms specifically designed for bioremediation is under preliminary research. Two category of genes can be inserted in the organism: degradative genes which encode proteins required for the degradation of pollutants, and reporter genes that are able to monitor pollution levels. Numerous members of *Pseudomonas* have also been modified with the lux gene, but for the detection of the polyaromatic hydrocarbon naphthalene. A field test for the release of the modified organism has been successful on a moderately large scale.

There are concerns surrounding release and containment of genetically modified organisms into the environment due to the potential of horizontal gene transfer. Genetically modified organisms are classified and controlled under the Toxic Substances Control Act of 1976 under United States Environmental Protection Agency. Measures have been created to address these concerns. Organisms can be modified such that they can only survive and grow under specific sets of environmental conditions. In addition, the tracking of modified organisms can be made easier with the insertion of bioluminescence genes for visual identification.

References

- Water-purification: britannica.com, Retrieved 22 January 2019

- "Chemists Work to Desalinate the Ocean for Drinking Water, One Nanoliter at a Time". Science Daily. June 27, 2013. Retrieved June 29, 2013

- Boiling, module-6-disaster-situations-planning-and-preparedness, further-resources-0: sswm.info, Retrieved 25 June 2019

- "Purification of Contaminated Water with Reverse Osmosis" ISSN 2250-2459, ISO 9001:2008 Certified Journal, Volume 3, Issue 12, December 2013

- Distillation-treatment-and-removal-contaminants-drinking-water: iwapublishing.com, Retrieved 14 April 2019

- MD-ZLD-interactive, Data-sheets: lenntech.com, Retrieved 10 August 2019

- Grabowski, Andrej (2010). Electromembrane desalination processes for production of low conductivity water. Berlin: Logos-Verl. ISBN 978-3832527143

5

Water Treatment Technologies

Water treatment refrs to the process that is used to improve the quality of water to make it potable and acceptable for industrial and agricultural activities. It makes use of various technologies such as coagulation, sedimentation, dissolved air flotation, disinfection, fluoridation, pH correction, etc. This chapter closely examines these technologies of water treatment to provide an extensive understanding of the subject.

Water Treatment

Water treatment is any process that improves the quality of water to make it more acceptable for a specific end-use. The end use may be drinking, industrial water supply, irrigation, river flow maintenance, water recreation or many other uses, including being safely returned to the environment. Water treatment removes contaminants and undesirable components, or reduces their concentration so that the water becomes fit for its desired end-use. This treatment is crucial to human health and allows humans to benefit from both drinking and irrigation use.

Dalecarlia Water Treatment Plant, Washington, D.C.

Treatment for Drinking Water Production

Treatment for drinking water production involves the removal of contaminants from raw water to produce water that is pure enough for human consumption without any short term or long term risk of any adverse health effect. In general terms, the greatest microbial risks are associated with ingestion of water that is contaminated with human or animal (including bird) faeces. Faeces can be a source of pathogenic bacteria, viruses, protozoa and helminths. [Guidelines for Drinking-water quality]. Substances that are removed during the process of drinking water treatment, Disinfection is of unquestionable importance in the supply of safe drinking-water. The destruction of microbial pathogens is essential and very commonly involves the use of reactive chemical agents such suspended solids, bacteria, algae, viruses, fungi, and minerals such as iron and manganese. These substances continue to cause great harm to several lower developed countries who do not have access to water purification.

The processes involved in removing the contaminants include physical processes such as settling and filtration, chemical processes such as disinfection and coagulation and biological processes such as slow sand filtration.

Measures taken to ensure water quality not only relate to the treatment of the water, but to its conveyance and distribution after treatment. It is therefore common practice to keep residual disinfectants in the treated water to kill bacteriological contamination during distribution.

World Health Organization (WHO) guidelines are a general set of standards intended to apply where better local standards are not implemented. More rigorous standards apply across Europe, the USA and in most other developed countries.

Water supplied to domestic properties, for tap water or other uses, may be further treated before use, often using an in-line treatment process. Such treatments can include water softening or ion exchange. Many proprietary systems also claim to remove residual disinfectants and heavy metal ions.

Processes

Empty aeration tank for iron precipitation.

Tanks with sand filters to remove precipitated iron (not working at the time).

A combination selected from the following processes is used for municipal drinking water treatment worldwide:

- Pre-chlorination for algae control and arresting biological growth.

- Aeration along with pre-chlorination for removal of dissolved iron when present with small amounts relatively of manganese.

- Coagulation for flocculation or slow-sand filtration.

- Coagulant aids, also known as polyelectrolytes – to improve coagulation and for more robust floc formation.

- Sedimentation for solids separation that is the removal of suspended solids trapped in the floc.

- Filtration to remove particles from water either by passage through a sand bed that can be washed and reused or by passage through a purpose designed filter that may be washable.

- Disinfection for killing bacteria, viruses and other pathogens.

Technologies for potable water and other uses are well developed, and generalized designs are available from which treatment processes can be selected for pilot testing on the specific source water. In addition, a number of private companies provide patented technological solutions for the treatment of specific contaminants. Automation of water treatment is common in the developed world. Source water quality through the seasons, scale, and environmental impact can dictate capital costs and operating costs. End use of the treated water dictates the necessary quality monitoring technologies, and locally available skills typically dictate the level of automation adopted.

Constituent	Unit Processes
Turbidity and particles	Coagulation/ flocculation, sedimentation, granular filtration
Major dissolved inorganics	Softening, aeration, membranes
Minor dissolved inorganics	Membranes
Pathogens	Sedimentation, filtration, disinfection
Major dissolved organics	Membranes, adsorption

Industrial Water Treatment

Two of the main processes of industrial water treatment are *boiler water treatment* and *cooling water treatment*. A large amount of proper water treatment can lead to the reaction of solids and bacteria within pipe work and boiler housing. Steam boilers can suffer from scale or corrosion when left untreated. Scale deposits can lead to weak and dangerous machinery, while additional fuel is required to heat the same level of water because of the rise in thermal resistance. Poor quality dirty water can become a breeding ground for bacteria such as *Legionella* causing a risk to public health.

Corrosion in low pressure boilers can be caused by dissolved oxygen, acidity and excessive alkalinity. Water treatment therefore should remove the dissolved oxygen and maintain the boiler water with the appropriate pH and alkalinity levels. Without effective water treatment, a cooling water system can suffer from scale formation, corrosion and fouling and may become a breeding ground for harmful bacteria. This reduces efficiency, shortens plant life and makes operations unreliable and unsafe.

Desalination

Saline water can be treated to yield fresh water. Two main processes are used, reverse osmosis or distillation. Both methods require more energy than water treatment of local surface waters, and are usually only used in coastal areas or where water such as groundwater has high salinity.

Portable Water Purification

Living away from drinking water supplies often requires some form of portable water treatment process. These can vary in complexity from the simple addition of a disinfectant tablet in a hiker's water bottle through to complex multi-stage processes carried by boat or plane to disaster areas.

Ultra Pure Water Production

Some industries such as the production of silicon wafers, space technology and many high quality metallurgical process require ultrapure water. The production of such water typically involves many stages, and can include reverse osmosis, ion exchange and several distillation stages using solid tin apparatus. This method is extremely useful by making water production extremely pure by the EPA water quality standards.

Developing Countries

Appropriate technology options in water treatment include both community-scale and household-scale point-of-use (POU) or self-supply designs. Such designs may employ solar water disinfection methods, using solar irradiation to inactivate harmful waterborne microorganisms directly, mainly by the UV-A component of the solar spectrum, or indirectly through the presence of an oxide photocatalyst, typically supported TiO_2 in its anatase or rutile phases. Despite progress in SODIS technology, military surplus water treatment units like the ERDLator are still frequently used in developing countries. Newer military style Reverse Osmosis Water Purification Units (ROWPU) are portable, self-contained water treatment plants are becoming more available for public use.

For waterborne disease reduction to last, water treatment programs that research and development groups start in developing countries must be sustainable by the citizens of those countries. This can ensure the efficiency of such programs after the departure of the research team, as monitoring is difficult because of the remoteness of many locations.

Energy Consumption

Water treatment plants can be significant consumers of energy. In California, more than 4% of the state's electricity consumption goes towards transporting moderate quality water over long distances, treating that water to a high standard. In areas with high quality water sources which flow by gravity to the point of consumption,, costs will be much lower. Much of the energy requirements are in pumping. Processes that avoid the need for pumping tend to have overall low energy demands. Those water treatment technologies that have very low energy requirements including trickling filters, slow sand filters, gravity aqueducts.

Pre-treatment

Pre-treatment is a critical part of the entire treatment plan when dealing with wastewater. How effectively it is implemented will dramatically affect primary treatment and impact how much maintenance is required for pre-treatment equipment.

Pre-treatment is most effectively done above ground. The primary reason for this is maintenance. Underground systems or equipment, while out of sight, are far more difficult to maintain and require confined space entry. Because there have been so many past incidents of workers becoming disabled in confined spaces, there are strict guidelines related to entry and operation in any confined space and it is always better to avoid creating them.

The first stage in pre-treatment is lifting wastewater above ground. The best way to accomplish this is with circular lift stations. Circular lift stations are self-cleaning, so they do not require pump-outs, as do in ground separators.

The second stage of pre-treatment is screening. Removing solids is an important step prior to any further treatment and is a big money saver. Many times the screened material can be handled either as compost or animal feed.

Once the wastewater is screened, solids settling is an option that can be effective if there is space and time. Many very fine and un-screenable solids will settle. An effective way to remove them is with cone bottom tanks. A cone bottom tank will concentrate sludge material into the bottom of the cone. It can then be removed to a storage tank, be sent to a dewatering hopper, or be sent to a filter press.

The final stage of pre-treatment is pH balancing. Further anaerobic or aerobic treatment requires a pH range that is close to neutral. Removing solids in the previous steps will dramatically reduce chemical consumption since solids act as sponges. The type of mixing in a pH system will also affect how much chemical is consumed.

Coagulation

In water treatment, coagulation flocculation involves the addition of compounds that promote the clumping of fines into larger floc so that they can be more easily separated from the water. Coagulation is a chemical process that involves neutralization of charge whereas flocculation is a physical process and does not involve neutralization of charge. The coagulation-flocculation process can be used as a preliminary or intermediary step between other water or wastewater treatment processes like filtration and sedimentation. Iron and aluminium salts are the most widely used coagulants but salts of other metals such as titanium and zirconium have been found to be highly effective as well.

Coagulation-flocculation process in a water treatment system.

Factors

Coagulation is affected by the type of coagulant used, its dose and mass; pH and initial turbidity of the water that is being treated; and properties of the pollutants present. The effectiveness of the coagulation process is also affected by pretreatments like oxidation.

Mechanism

In a colloidal suspension, particles will settle very slowly or not at all because the colloidal particles carry surface electrical charges that mutually repel each other. This surface charge is most commonly evaluated in terms of zeta potential, the electrical potential at the slipping plane. To induce coagulation, a coagulant (typically a metallic salt) with the opposite charge is added to the water to overcome the repulsive charge and "destabilize" the suspension. For example, the colloidal particles are negatively charged and alum is added as a coagulant to create positively charged ions. Once the repulsive charges have been neutralized (since opposite charges attract), van der Waals force will cause the particles to cling together (agglomerate) and form micro floc.

Determining Coagulant Dose

Jar Test

The dose of the coagulant to be used can be determined via the jar test. The jar test involves exposing same volume samples of the water to be treated to different doses of the coagulant and then simultaneously mixing the samples at a constant rapid mixing time. The microfloc formed after coagulation further undergoes flocculation and is allowed to settle. Then the turbidity of the samples is measured and the dose with the lowest turbidity can be said to be optimum.

Jar test for coagulation.

Microscale Dewatering Tests

Despite its widespread use in the performance of so-called "dewatering experiments", the jar test is limited in its usefulness due to several disadvantages. For example, evaluating the performance of prospective coagulants or flocculants requires both significant volumes of water/wastewater samples (liters) and experimental time (hours). This limits the scope of the experiments which can be conducted, including the addition of replicates. Furthermore, the analysis of jar test experiments produces results which are often only semi-quantitative. Coupled with the wide range of chemical coagulants and flocculants that exist, it has been remarked that determining the most appropriate dewatering agent as well as the optimal dose "is widely considered to be more of an 'art' rather than a 'science'". As such, dewatering performance tests such as the jar test lend themselves well to miniaturization. For example, the Microscale Flocculation Test developed by LaRue *et al.* reduces the scale of conventional jar tests down to the size of a standard multi-well microplate, which yields benefits stemming from the reduced sample volume and increased parallelization; this technique is also amenable to quantitative dewatering metrics, such as capillary suction time.

Streaming Current Detector

An automated device for determining the coagulant dose is the Streaming Current Detector (SCD). The SCD measures the net surface charge of the particles and shows a streaming current value of 0 when the charges are neutralized (cationic coagulants neutralize the anionic colloids). At this value (0), the coagulant dose can be said to be optimum.

Limitations

Coagulation itself results in the formation of floc but flocculation is required to help the floc further aggregate and settle. The coagulation-flocculation process itself removes only about 60%-70% of

Natural Organic Matter (NOM) and thus, other processes like oxidation, filtration and sedimentation are necessary for complete raw water or wastewater treatment. Coagulant aids (polymers that bridge the colloids together) are also often used to increase the efficiency of the process.

Sedimentation

Sedimentation is a physical water treatment process using gravity to remove suspended solids from water. Solid particles entrained by the turbulence of moving water may be removed naturally by sedimentation in the still water of lakes and oceans. Settling basins are ponds constructed for the purpose of removing entrained solids by sedimentation. Clarifiers are tanks built with mechanical means for continuous removal of solids being deposited by sedimentation. This can also be seen, for example in Aromatherapy oils. Clarification does not remove dissolved species. Sedimentation is the act of depositing sediment.

Basics

Suspended solids (or SS), is the mass of dry solids retained by a filter of a given porosity related to the volume of the water sample. This includes particles 10 μm and greater.

Colloids are particles of a size between 1 nm (0.001 μm) and 1 μm depending on the method of quantification. Because of Brownian motion and electrostatic forces balancing the gravity, they are not likely to settle naturally.

The limit sedimentation velocity of a particle is its theoretical descending speed in clear and still water. In settling process theory, a particle will settle only if :-

- In a vertical ascending flow, the ascending water velocity is lower than the limit sedimentation velocity.

- In a longitudinal flow, the ratio of the length of the tank to the height of the tank is higher than the ratio of the water velocity to the limit sedimentation velocity.

Removal of suspended particles by sedimentation depends upon the size, zeta potential and specific gravity of those particles. Suspended solids retained on a filter may remain in suspension if their specific gravity is similar to water while very dense particles passing through the filter may settle. Settleable solids are measured as the visible volume accumulated at the bottom of an Imhoff cone after water has settled for one hour.

Gravitational theory is employed, alongside the derivation from Newton's second law and the Navier-Stokes equations.

Stokes' law explains the relationship between the settling rate and the particle diameter. Under specific conditions, the particle settling rate is directly proportional to the square of particle diameter and inversely proportional to liquid viscosity.

The settling velocity, defined as the residence time taken for the particles to settle in the tank, enables the calculation of tank volume. Precise design and operation of a sedimentation tank is of high importance in order to keep the amount of sediment entering the diversion system to a minimum threshold by maintaining the transport system and stream stability to remove the sediment diverted from the system. This is achieved by reducing stream velocity as low as possible for the longest period of time possible. This is feasible by widening the approach channel and lowering its floor to reduce flow velocity thus allowing sediment to settle out of suspension due to gravity. The settling behavior of heavier particulates is also affected by the turbulence.

Designs

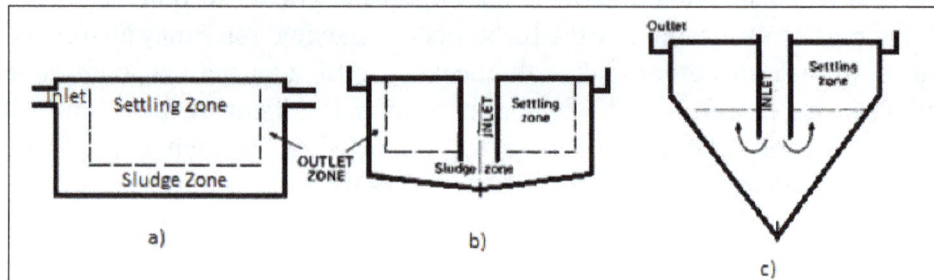

Different clarifier designs.

Although sedimentation might occur in tanks of other shapes, removal of accumulated solids is easiest with conveyor belts in rectangular tanks or with scrapers rotating around the central axis of circular tanks. Settling basins and clarifiers should be designed based on the settling velocity of the smallest particle to be theoretically 100% removed. The overflow rate is defined as:

- Overflow rate (V_0) = Flow of water (Q (cubic metre per second)) /(Surface area of settling basin (A))(m^2).

- In many countries this value is named as surface loading in m³/h per m². Overflow rate is often used for flow over an edge (for example a weir) in the unit m³/h per m.

The unit of overflow rate is usually meters (or feet) per second, a velocity. Any particle with settling velocity (*Vs*) greater than the overflow rate will settle out, while other particles will settle in the ratio Vs/Vo. There are recommendations on the overflow rates for each design that ideally take into account the change in particle size as the solids move through the operation:

- Quiescent zones: 9.4 mm (0.031 ft) per second.

- Full-flow basins: 4.0 mm (0.013 ft) per second.

- Off-line basins: 0.46 mm (0.0015 ft) per second.

However, factors such as flow surges, wind shear, scour, and turbulence reduce the effectiveness of settling. To compensate for these less than ideal conditions, it is recommended doubling the area calculated by the previous equation. It is also important to equalize flow distribution at each point across the cross-section of the basin. Poor inlet and outlet designs can produce extremely poor flow characteristics for sedimentation.

Settling basins and clarifiers can be designed as long rectangles, that are hydraulically more stable and easier to control for large volumes. Circular clarifiers work as a common thickener (without the usage of rakes), or as upflow tanks.

Sedimentation efficiency does not depend on the tank depth. If the forward velocity is low enough so that the settled material does not re-suspend from the tank floor, the area is still the main parameter when designing a settling basin or clarifier, taking care that the depth is not too low.

Assessment of Main Process Characteristics

Settling basins and clarifiers are designed to retain water so that suspended solids can settle. By sedimentation principles, the suitable treatment technologies should be chosen depending on the specific gravity, size and shear resistance of particles. Depending on the size and density of particles, and physical properties of the solids, there are four types of sedimentation processes:

- Type 1 – Dilutes, non-flocculent, free-settling (every particle settles independently).

- Type 2 – Dilute, flocculent (particles can flocculate as they settle).

- Type 3 – Concentrated suspensions, zone settling, hindered settling (sludge thickening).

- Type 4 – Concentrated suspensions, compression (sludge thickening).

Different factors control the sedimentation rate in each.

Settling of Discrete Particles

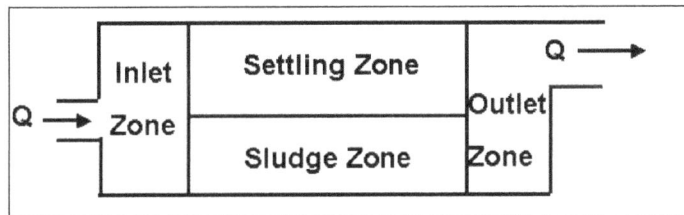

The four functional zones of a continuous flow settling basin.

Unhindered settling is a process that removes the discrete particles in a very low concentration without interference from nearby particles. In general, if the concentration of the solutions is lower than 500 mg/L total suspended solids, sedimentation will be considered discrete. Concentrations of raceway effluent total suspended solids (TSS) in the west are usually less than 5 mg/L net. TSS concentrations of off-line settling basin effluent are less than 100 mg/L net. The particles keep their size and shape during discrete settling, with an independent velocity. With such low concentrations of suspended particles, the probability of particle collisions is very low and consequently the rate of flocculation is small enough to be neglected for most calculations. Thus the surface area of the settling basin becomes the main factor of sedimentation rate. All continuous flow settling basins are divided into four parts: inlet zone, settling zone, sludge zone and outlet zone.

In the inlet zone, flow is established in a same forward direction. Sedimentation occurs in the settling zone as the water flow towards to outlet zone. The clarified liquid is then flow out from outlet zone. Sludge zone: settled will be collected here and usually we assume that it is removed from water flow once the particles arrives the sludge zone.

In an ideal rectangular sedimentation tank, in the settling zone, the critical particle enters at the top of the settling zone, and the settle velocity would be the smallest value to reach the sludge zone, and at the end of outlet zone, the velocity component of this critical particle are Vs, the settling velocity in vertical direction and Vh in horizontal direction.

From figure, the time needed for the particle to settle;

$$t\_o = H/V\_p = L/Vs$$

Since the surface area of the tank is WL, and Vs = Q/WL, Vh = Q/WH, where Q is the flow rate and W, L, H is the width, length, depth of the tank.

According to equation, this also is a basic factor that can control the sedimentation tank performance which called overflow rate.

Eq. also tell us that the depth of sedimentation tank is independent to the sedimentation efficiency, only if the forward velocity is low enough to make sure the settled mass would not suspended again from the tank floor.

Settlement of Flocculent Particles

In a horizontal sedimentation tank, some particles may not follow the diagonal line in Fig., while settling faster as they grow. So this says that particles can grow and develop a higher settling velocity if a greater depth with longer retention time. However, the collision chance would be even greater if the same retention time were spread over a longer, shallower tank. In fact, in order to avoid hydraulic short-circuiting, tanks usually are made 3–6 m deep with retention times of a few hours.

Zone-settling Behaviour

As the concentration of particles in a suspension is increased, a point is reached where particles are so close together that they no longer settle independently of one another and the velocity fields of the fluid displaced by adjacent particles, overlap. There is also a net upward flow of liquid displaced by the settling particles. This results in a reduced particle-settling velocity and the effect is known as hindered settling.

There is a common case for hindered settling occurs. the whole suspension tends to settle as a 'blanket' due to its extremely high particle concentration. This is known as zone settling, because it is easy to make a distinction between several different zones which separated by concentration discontinuities. Fig. represents a typical batch-settling column tests on a suspension exhibiting zone-settling characteristics. There is a clear interface near the top of the column would be formed to separating the settling sludge mass from the clarified supernatant as long as leaving such a suspension to stand in a settling column. As the suspension settles, this interface will move down at the same speed. At the same time, there is an interface near the bottom between that settled suspension and the suspended blanket. After settling of suspension is complete, the bottom interface would move upwards and meet the top interface which moves downwards.

Compression Settling

The settling particles can contact each other and arise when approaching the floor of the sedimentation tanks at very high particle concentration. So that further settling will only occur in adjust matrix as the sedimentation rate decreasing. This is can be illustrated by the lower region of the zone-settling diagram. In Compression zone, the settled solids are compressed by gravity (the weight of solids), as the settled solids are compressed under the weight of overlying solids, and water is squeezed out while the space gets smaller.

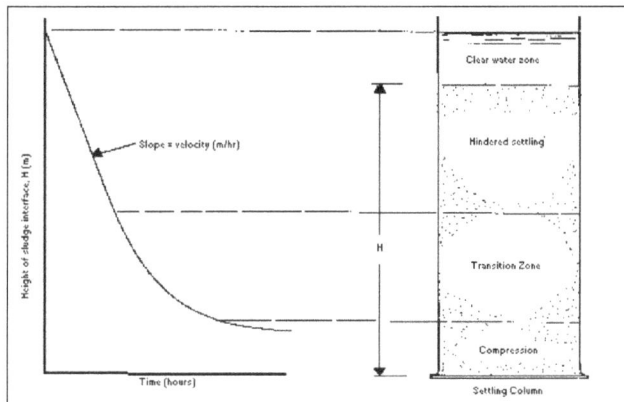

Typical batch-settling column test on a suspension exhibiting zone-settling characteristics.

Applications

Potable Water Treatment

Sedimentation in potable water treatment generally follows a step of chemical coagulation and flocculation, which allows grouping particles together into flocs of a bigger size. This increases the settling speed of suspended solids and allows settling colloids.

Wastewater Treatment

Sedimentation has been used to treat wastewater for millennia.

Primary treatment of sewage is removal of floating and settleable solids through sedimentation. *Primary clarifiers* reduce the content of suspended solids as well as the pollutant embedded in the suspended solids. Because of the large amount of reagent necessary to treat domestic wastewater, preliminary chemical coagulation and flocculation are generally not used, remaining suspended solids being reduced by following stages of the system. However, coagulation and flocculation can be used for building a compact treatment plant (also called a "package treatment plant"), or for further polishing of the treated water.

Sedimentation tanks called "secondary clarifiers" remove flocs of biological growth created in some methods of secondary treatment including activated sludge, trickling filters and rotating biological contactors.

Dissolved Air Flotation

Dissolved air flotation (DAF) is a water treatment process that clarifies wastewaters (or other waters) by the removal of suspended matter such as oil or solids. The removal is achieved by dissolving air in the water or wastewater under pressure and then releasing the air at atmospheric pressure in a flotation tank basin. The released air forms tiny bubbles which adhere to the suspended matter causing the suspended matter to float to the surface of the water where it may then be removed by a skimming device.

Dissolved air flotation is very widely used in treating the industrial wastewater effluents from oil refineries, petrochemical and chemical plants, natural gas processing plants, paper mills, general water treatment and similar industrial facilities. A very similar process known as *induced gas flotation* is also used for wastewater treatment. *Froth flotation* is commonly used in the processing of mineral ores.

In the oil industry, *dissolved gas flotation* (DGF) units do not use air as the flotation medium due to the explosion risk. Nitrogen gas is used instead to create the bubbles.

Process Description

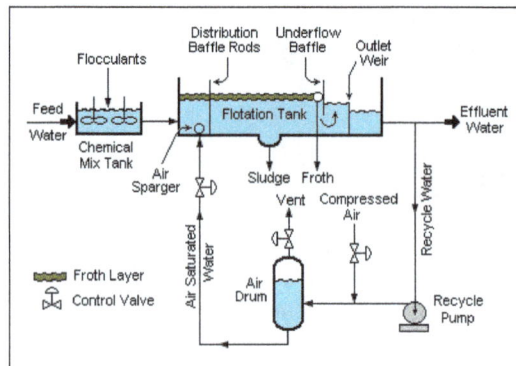

A typical dissolved air flotation unit (DAF).

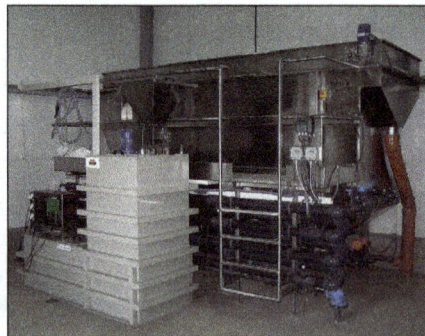

DAF unit with a capacity of 20 m³/h, visible also: flocculant preparation station and pipe flocculator.

Modern DAF units using parallel plate technology are quite compact.
Picture shows a 225 m³/h DAF.

The feed water to the DAF float tank is often (but not always) dosed with a coagulant (such as ferric chloride or aluminum sulfate) to coagulate the colloidal particles and/or a flocculant to conglomerate the particles into bigger clusters.

A portion of the clarified effluent water leaving the DAF tank is pumped into a small pressure vessel (called the air drum) into which compressed air is also introduced. This results in saturating the pressurized effluent water with air. The air-saturated water stream is recycled to the front of the float tank and flows through a pressure reduction valve just as it enters the front of the float tank, which results in the air being released in the form of tiny bubbles. Bubbles form at nucleation sites on the surface of the suspended particles, adhering to the particles. As more bubbles form, the lift from the bubbles eventually overcomes the force of gravity. This causes the suspended matter to float to the surface where it forms a froth layer which is then removed by a skimmer. The froth-free water exits the float tank as the clarified effluent from the DAF unit.

Some DAF unit designs utilize parallel plate packing material (e.g. lamellas) to provide more separation surface and therefore to enhance the separation efficiency of the unit.

DAF systems can be categorized as circular (more efficient) and rectangular (more residence time). The former type requires just 3 minutes. A particular circular DAF system is called "Zero speed", allowing quite water status then highest performances. The rectangular type requires 20 to 30 minutes. One of the bigger advantages of the circular type is its spiral scoop.

Drinking Water Treatment

Drinking water supplies that are particularly vulnerable to unicellular algal blooms, and supplies with low turbidity and high colour often employ DAF. After coagulation and flocculation processes, water flows to DAF tanks where air diffusers on the tank bottom create fine bubbles that attach to floc resulting in a floating mass of concentrated floc. The floating floc blanket is removed from the surface and clarified water is withdrawn from the bottom of the DAF tank.

Removal of Ions and other Dissolved Substances

Ultrafiltration membranes use polymer membranes with chemically formed microscopic pores that can be used to filter out dissolved substances avoiding the use of coagulants. The type of membrane media determines how much pressure is needed to drive the water through and what sizes of micro-organisms can be filtered out.

Ion exchange: Ion exchange systems use ion exchange resin- or zeolite-packed columns to replace unwanted ions. The most common case is water softening consisting of removal of Ca^{2+} and Mg^{2+} ions replacing them with benign (soap friendly) Na^+ or K^+ ions. Ion exchange resins are also used to remove toxic ions such as nitrite, lead, mercury, arsenic and many others.

Precipitative softening: Water rich in hardness (calcium and magnesium ions) is treated with lime (calcium oxide) and/or soda-ash (sodium carbonate) to precipitate calcium carbonate out of solution utilizing the common-ion effect.

Electrodeionization: Water is passed between a positive electrode and a negative electrode. Ion exchange membranes allow only positive ions to migrate from the treated water toward the negative electrode and only negative ions toward the positive electrode. High purity deionized water is produced continuously, similar to ion exchange treatment. Complete removal of ions from water is possible if the right conditions are met. The water is normally pre-treated with a reverse osmosis unit to remove non-ionic organic contaminants, and with gas transfer membranes to remove carbon dioxide. A water recovery of 99% is possible if the concentrate stream is fed to the RO inlet.

Disinfection

Water disinfection means the removal, deactivation or killing of pathogenic microorganisms. Microorganisms are destroyed or deactivated, resulting in termination of growth and reproduction. When microorganisms are not removed from drinking water, drinking water usage will cause people to fall ill.

Sterilization is a process related to disinfection. However, during the sterilization process all present microorganisms are killed, both harmful and harmless microorganisms.

Media

Disinfection can be attained by means of physical or chemical disinfectants. The agents also remove organic contaminants from water, which serve as nutrients or shelters for microorganisms. Disinfectants should not only kill microorganisms. Disinfectants must also have a residual effect, which means that they remain active in the water after disinfection. A disinfectant should prevent pathogenic microorganisms from growing in the plumbing after disinfection, causing the water te be recontaminated.

For chemical disinfection of water the following disinfectants can be used:

- Chlorine (Cl_2),
- Chlorine dioxide (ClO_2),
- Hypo chlorite (OCl^-),
- Ozone (O_3),
- Halogens: bromine (Br_2), iodene (I),
- Bromine chloride ($BrCl$),
- Metals: copper (Cu^{2+}), silver (Ag^+),
- Kaliumpermanganate ($KMnO_4$),
- Fenols,
- Alcohols,

- Soaps and detergents,

- Kwartair ammonium salts,

- Hydrogen peroxide,

- Several acids and bases.

For physical disinfection of water the following disinfectants can be used:

- Ultraviolet light (UV),

- Electronic radiation,

- Gamma rays,

- Sounds,

- Heat.

Chemical inactivation of microbiological contamination in natural or untreated water is usually one of the final steps to reduce pathogenic microorganisms in drinking water. Combinations of water purification steps (oxidation, coagulation, settling, disinfection, filtration) cause (drinking) water to be safe after production. As an extra measure many countries apply a second disinfection step at the end of the water purification process, in order to protect the water from microbiological contamination in the water distribution system. Usually one uses a different kind of disinfectant from the one earlier in the process, during this disinfection process. The secundairy disinfection makes sure that bacteria will not multiply in the water during distribution. Bacteria can remain in the water after the first disinfection step or can end up in the water during backflushing of contaminated water (which can contain groundwater bacteria as a result of cracks in the plumbing).

Disinfection Mechanism

Disinfection commonly takes place because of cell wall corrosion in the cells of microorganisms, or changes in cell permeability, protoplasm or enzyme activity (because of a structural change in enzymes). These disturbances in cell activity cause microorganisms to no longer be able to multiply. This will cause the microorganisms to die out. Oxidizing disinfectants also demolish organic matter in the water, causing a lack of nutrients.

Fluoridation

Water fluoridation is the controlled adjustment of fluoride to a public water supply to reduce tooth decay. Fluoridated water contains fluoride at a level that is effective for preventing cavities; this can occur naturally or by adding fluoride. Fluoridated water operates on tooth surfaces: in the mouth, it creates low levels of fluoride in saliva, which reduces the rate at which tooth enamel demineralizes and increases the rate at which it remineralizes in the early stages of cavities. Typically a fluoridated compound is added to drinking water, a process that in the U.S. costs an average of about $1.08 per person-year. Defluoridation is needed when the naturally occurring fluoride level exceeds recommended limits. In 2011 the World Health Organization suggested a level of fluoride from 0.5 to 1.5 mg/L (milligrams per litre), depending on climate, local environment, and other sources of fluoride. Bottled water typically has unknown fluoride levels.

Tooth decay remains a major public health concern in most industrialized countries, affecting 60–90% of schoolchildren and the vast majority of adults. Water fluoridation reduces cavities in children, while efficacy in adults is less clear. A Cochrane review estimates a reduction in cavities when water fluoridation was used by children who had no access to other sources of fluoride to be 35% in baby teeth and 26% in permanent teeth. However this was based on older studies which failed to control for numerous variables, such as increasing sugar consumption as well as other dental strategies. Most European countries have experienced substantial declines in tooth decay, though milk and salt fluoridation is widespread in lieu of water fluoridation. Recent studies suggest that water fluoridation, particularly in industrialized nations, may be unnecessary because topical fluorides (such as in toothpaste) are widely used, and caries rates have become low.

Although fluoridation can cause dental fluorosis, which can alter the appearance of developing teeth or enamel fluorosis, the differences are mild and usually not an aesthetic or public health concern. There is no clear evidence of other adverse effects from water fluoridation. Fluoride's effects depend on the total daily intake of fluoride from all sources. Drinking water is typically the largest source; other methods of fluoride therapy include fluoridation of toothpaste, salt, and milk. The views on the most efficient method for community prevention of tooth decay are mixed. The Australian government states that water fluoridation is the most effective way to achieve fluoride exposure that is community-wide. The World Health Organization reports that water fluoridation, when feasible and culturally acceptable, has substantial advantages, especially for subgroups at high risk, while the European Commission finds no benefit to water fluoridation compared with topical use.

Fluoridation does not affect the appearance, taste or smell of drinking water.

Public water fluoridation was first practiced in the U.S. As of 2012, 25 countries have artificial water fluoridation to varying degrees, 11 of them have more than 50% of their population drinking fluoridated water. A further 28 countries have water that is naturally fluoridated, though in many of them the fluoride is above the optimal level. As of 2012, about 435 million people worldwide received water fluoridated at the recommended level (i.e., about 5.4% of the global population).

About 214 million of them living in the United States. Major health organizations such as the World Health Organization and FDI World Dental Federation supported water fluoridation as safe and effective. The Centers for Disease Control and Prevention lists water fluoridation as one of the ten great public health achievements of the 20th century in the U.S. Despite this, the practice is controversial as a public health measure. Some countries and communities have discontinued fluoridation, while others have expanded it. Opponents of the practice argue that neither the benefits nor the risks have been studied adequately, and debate the conflict between what might be considered mass medication and individual liberties.

Goal

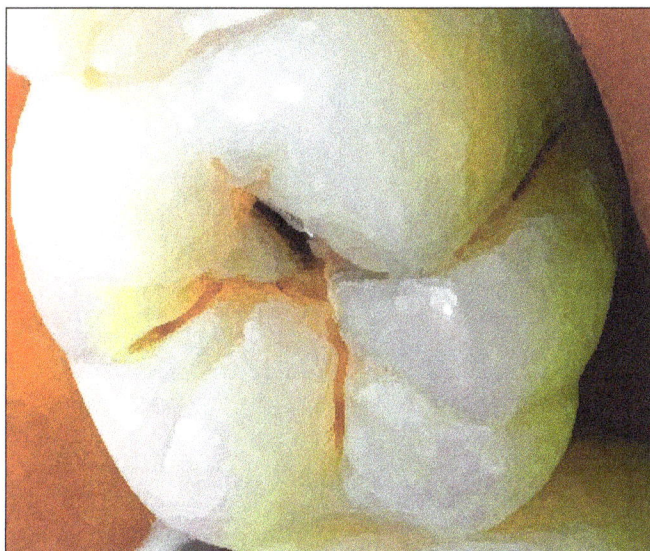

A cavity starts in a tooth's outer enamel and spreads to the dentin and pulp inside.

The goal of water fluoridation is to prevent tooth decay by adjusting the concentration of fluoride in public water supplies. Tooth decay (dental caries) is one of the most prevalent chronic diseases worldwide. Although it is rarely life-threatening, tooth decay can cause pain and impair eating, speaking, facial appearance, and acceptance into society, and it greatly affects the quality of life of children, particularly those of low socioeconomic status. In most industrialized countries, tooth decay affects 60–90% of schoolchildren and the vast majority of adults; although the problem appears to be less in Africa's developing countries, it is expected to increase in several countries there because of changing diet and inadequate fluoride exposure. In the U.S., minorities and the poor both have higher rates of decayed and missing teeth, and their children have less dental care. Once a cavity occurs, the tooth's fate is that of repeated restorations, with estimates for the median life of an amalgam tooth filling ranging from 9 to 14 years. Oral disease is the fourth most expensive disease to treat. The motivation for fluoridation of salt or water is similar to that of iodized salt for the prevention of congenital hypothyroidism and goiter.

The goal of water fluoridation is to prevent a chronic disease whose burdens particularly fall on children and the poor. Another of the goals was to bridge inequalities in dental health and dental care. Some studies suggest that fluoridation reduces oral health inequalities between the rich and poor, but the evidence is limited. There is anecdotal but not scientific evidence that fluoride allows more time for dental treatment by slowing the progression of tooth decay, and that it simplifies treatment by causing most cavities to occur in pits and fissures of teeth. Other reviews have found not enough evidence to determine if water fluoridation reduces oral-health social disparities.

Health and dental organizations worldwide have endorsed its safety and effectiveness. Its use began in 1945, following studies of children in a region where higher levels of fluoride occur naturally in the water. Further research showed that moderate fluoridation prevents tooth decay.

Implementation

Fluoride monitor (at left) in a community water tower pumphouse.

Fluoridation does not affect the appearance, taste, or smell of drinking water. It is normally accomplished by adding one of three compounds to the water: sodium fluoride, fluorosilicic acid, or sodium fluorosilicate.

- Sodium fluoride (NaF) was the first compound used and is the reference standard. It is a white, odorless powder or crystal; the crystalline form is preferred if manual handling is used, as it minimizes dust. It is more expensive than the other compounds, but is easily handled and is usually used by smaller utility companies. It is toxic in gram quantities by ingestion or inhalation.

- Fluorosilicic acid (H_2SiF_6) is the most commonly used additive for water fluoridation in the United States. It is an inexpensive liquid by-product of phosphate fertilizer manufacture. It comes in varying strengths, typically 23–25%; because it contains so much water, shipping can be expensive. It is also known as hexafluorosilicic, hexafluosilicic, hydrofluosilicic, and silicofluoric acid.

- Sodium fluorosilicate (Na_2SiF_6) is the sodium salt of fluorosilicic acid. It is a powder or very fine crystal that is easier to ship than fluorosilicic acid. It is also known as sodium silicofluoride.

These compounds were chosen for their solubility, safety, availability, and low cost. A 1992 census found that, for U.S. public water supply systems reporting the type of compound used, 63% of the population received water fluoridated with fluorosilicic acid, 28% with sodium fluorosilicate, and 9% with sodium fluoride.

Recommendations

The Centers for Disease Control and Prevention developed recommendations for water fluoridation that specify requirements for personnel, reporting, training, inspection, monitoring, surveillance, and actions in case of overfeed, along with technical requirements for each major compound used.

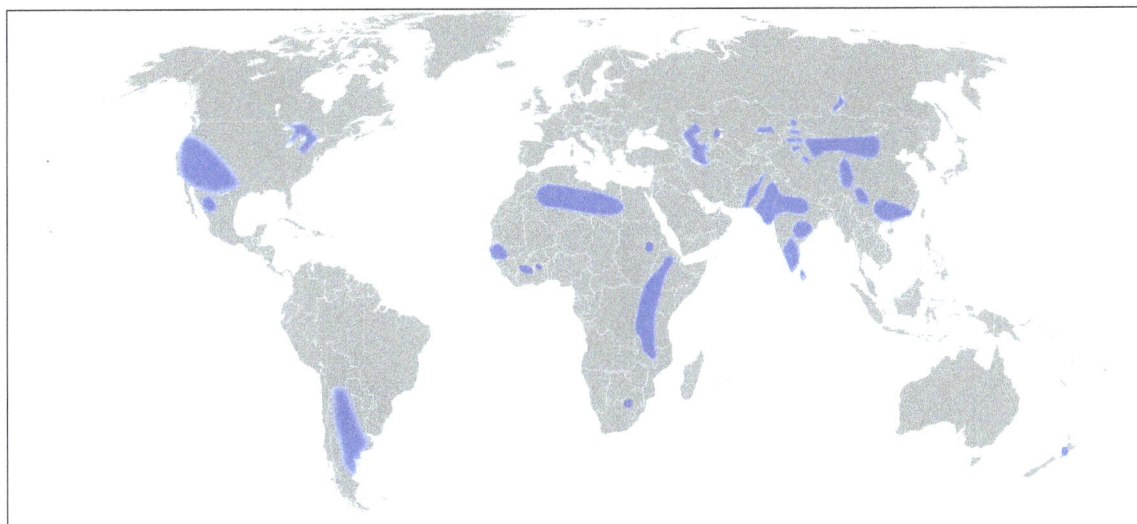

Geographical areas associated with groundwater having over 1.5 mg/L
of naturally occurring fluoride, which is above recommended levels.

Although fluoride was once considered an essential nutrient, the U.S. National Research Council has since removed this designation due to the lack of studies showing it is essential for human growth, though still considering fluoride a "beneficial element" due to its positive impact on oral health. The European Food Safety Authority's Panel on Dietetic Products, Nutrition and Allergies (NDA) considers fluoride not to be an essential nutrient, yet, due to the beneficial effects of dietary fluoride on prevention of dental caries they have defined an Adequate Intake (AI) value for it. The AI of fluoride from all sources (including non-dietary sources) is 0.05 mg/kg body weight per day for both children and adults, including pregnant and lactating women.

In 2011, the U.S. Department of Health and Human Services (HHS) and the U.S. Environmental Protection Agency (EPA) lowered the recommended level of fluoride to 0.7 mg/L. In 2015, the U.S. Food and Drug Administration (FDA), based on the recommendation of the U.S. Public Health Service (PHS) for fluoridation of community water systems, recommended that bottled water manufacturers limit fluoride in bottled water to no more than 0.7 milligrams per liter (mg/L) (milligrams per liter, equivalent to parts per million).

Previous recommendations were based on evaluations from 1962, when the U.S. specified the optimal level of fluoride to range from 0.7 to 1.2 mg/L (milligrams per liter, equivalent to parts per million), depending on the average maximum daily air temperature; the optimal level is lower in warmer climates, where people drink more water, and is higher in cooler climates.

These standards are not appropriate for all parts of the world, where fluoride levels might be excessive and fluoride should be removed from water, and is based on assumptions that have become obsolete with the rise of air conditioning and increased use of soft drinks, processed food, fluoridated toothpaste, and other sources of fluorides. In 2011 the World Health Organization stated that 1.5 mg/L should be an absolute upper bound and that 0.5 mg/L may be an appropriate lower limit. A 2007 Australian systematic review recommended a range from 0.6 to 1.1 mg/L.

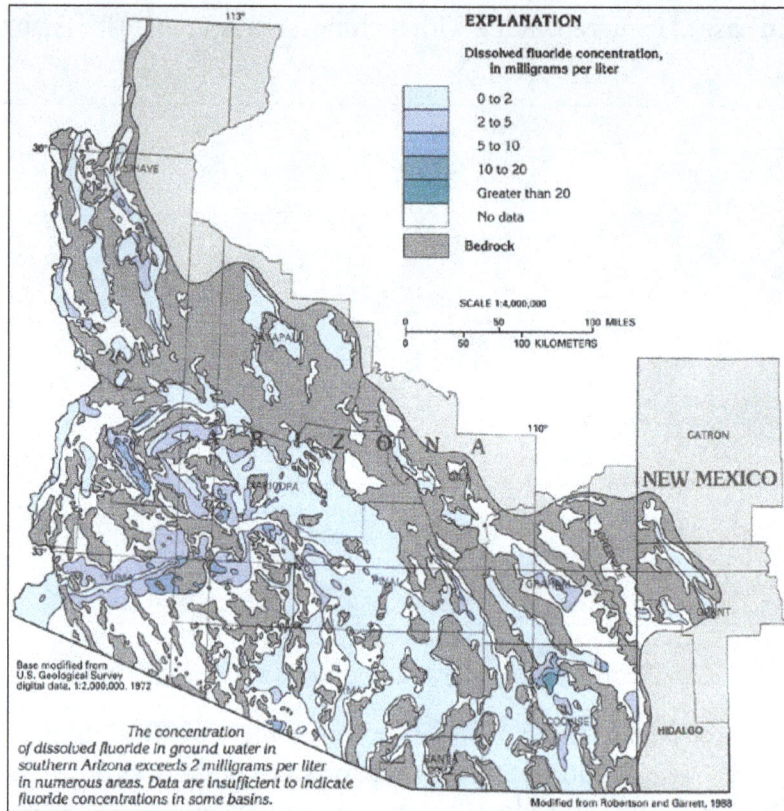

Detail of southern Arizona. Areas in darker blues have groundwater with over 2 mg/L of naturally occurring fluoride.

Occurrences

Fluoride naturally occurring in water can be above, at, or below recommended levels. Rivers and lakes generally contain fluoride levels less than 0.5 mg/L, but groundwater, particularly in volcanic or mountainous areas, can contain as much as 50 mg/L. Higher concentrations of fluorine are found in alkaline volcanic, hydrothermal, sedimentary, and other rocks derived from highly evolved magmas and hydrothermal solutions, and this fluorine dissolves into nearby water as fluoride. In most drinking waters, over 95% of total fluoride is the F^- ion, with the magnesium–fluoride complex (MgF^+) being the next most common. Because fluoride levels in water are usually controlled by the solubility of fluorite (CaF_2), high natural fluoride levels are associated with calcium-deficient, alkaline, and soft waters. Defluoridation is needed when the naturally occurring fluoride level exceeds recommended limits. It can be accomplished by percolating water through granular beds of activated alumina, bone meal, bone char, or tricalcium phosphate; by coagulation with alum; or by precipitation with lime.

Pitcher or faucet-mounted water filters do not alter fluoride content; the more-expensive reverse osmosis filters remove 65–95% of fluoride, and distillation removes all fluoride. Some bottled waters contain undeclared fluoride, which can be present naturally in source waters, or if water is sourced from a public supply which has been fluoridated. The FDA states that bottled water products labeled as de-ionized, purified, demineralized, or distilled have been treated in such a way that they contain no or only trace amounts of fluoride, unless they specifically list fluoride as an added ingredient.

Evidence

Existing evidence suggests that water fluoridation reduces tooth decay. Consistent evidence also suggests that it causes dental fluorosis, most of which is mild and not usually of aesthetic concern. No clear evidence of other adverse effects exists, though almost all research thereof has been of poor quality.

Effectiveness

Reviews have shown that water fluoridation reduces cavities in children. A conclusion for the efficacy in adults is less clear with some reviews finding benefit and others not. Studies in the U.S. in the 1950s and 1960s showed that water fluoridation reduced childhood cavities by fifty to sixty percent, while studies in 1989 and 1990 showed lower reductions (40% and 18% respectively), likely due to increasing use of fluoride from other sources, notably toothpaste, and also the 'halo effect' of food and drink that is made in fluoridated areas and consumed in unfluoridated ones.

A 2000 UK systematic review (York) found that water fluoridation was associated with a decreased proportion of children with cavities of 15% and with a decrease in decayed, missing, and filled primary teeth (average decreases was 2.25 teeth). The review found that the evidence was of moderate quality: few studies attempted to reduce observer bias, control for confounding factors, report variance measures, or use appropriate analysis. Although no major differences between natural and artificial fluoridation were apparent, the evidence was inadequate for a conclusion about any differences. A 2007 Australian systematic review used the same inclusion criteria as York's, plus one additional study. This did not affect the York conclusions. A 2011 European Commission systematic review based its efficacy on York's review conclusion. A 2015 Cochrane systematic review estimated a reduction in cavities when water fluoridation was used by children who had no access to other sources of fluoride to be 35% in baby teeth and 26% in permanent teeth. The evidence was of poor quality.

Fluoride may also prevent cavities in adults of all ages. A 2007 meta-analysis by CDC researchers found that water fluoridation prevented an estimated 27% of cavities in adults, about the same fraction as prevented by exposure to any delivery method of fluoride (29% average). A 2011 European Commission review found that the benefits of water fluoridation for adult in terms of reductions in decay are limited. A 2015 Cochrane review found no conclusive research regarding the effectiveness of water fluoridation in adults. A 2016 review found variable quality evidence that, overall, stopping of community water fluoridation programs was typically followed by an increase in cavities.

Most countries in Europe have experienced substantial declines in cavities without the use of water fluoridation due to the introduction of fluoridated toothpaste and the large use of other fluoride-containing products, including mouthrinse, dietary supplements, and professionally applied or prescribed gel, foam, or varnish. For example, in Finland and Germany, tooth decay rates remained stable or continued to decline after water fluoridation stopped in communities with widespread fluoride exposure from other sources. Fluoridation is however still clearly necessary in the U.S. because unlike most European countries, the U.S. does not have school-based dental care, many children do not visit a dentist regularly, and for many U.S. children water fluoridation is the prime source of exposure to fluoride. The effectiveness of water fluoridation can vary according to circumstances such as whether preventive dental care is free to all children.

Fluorosis

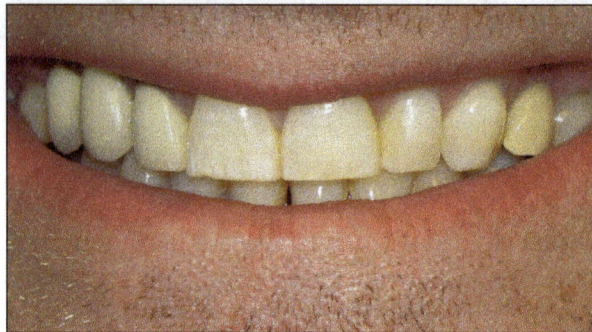

A mild case of dental fluorosis, visible as white streaks on the subject's upper right central incisor.

Fluoride's adverse effects depend on total fluoride dosage from all sources. At the commonly recommended dosage, the only clear adverse effect is dental fluorosis, which can alter the appearance of children's teeth during tooth development; this is mostly mild and is unlikely to represent any real effect on aesthetic appearance or on public health. In April 2015, recommended fluoride levels in the United States were changed to 0.7 ppm from 0.7–1.2 ppm to reduce the risk of dental fluorosis. The 2015 Cochrane review estimated that for a fluoride level of 0.7 ppm the percentage of participants with fluorosis of aesthetic concern was approximately 12%. This increases to 40% when considering fluorosis of any level not of aesthetic concern. In the US mild or very mild dental fluorosis has been reported in 20% of the population, moderate fluorosis in 2% and severe fluorosis in less than 1%.

The critical period of exposure is between ages one and four years, with the risk ending around age eight. Fluorosis can be prevented by monitoring all sources of fluoride, with fluoridated water directly or indirectly responsible for an estimated 40% of risk and other sources, notably toothpaste, responsible for the remaining 60%. Compared to water naturally fluoridated at 0.4 mg/L, fluoridation to 1 mg/L is estimated to cause additional fluorosis in one of every 6 people (95% CI 4–21 people), and to cause additional fluorosis of aesthetic concern in one of every 22 people (95% CI 13.6–∞ people). Here, *aesthetic concern* is a term used in a standardized scale based on what adolescents would find unacceptable, as measured by a 1996 study of British 14-year-olds. In many industrialized countries the prevalence of fluorosis is increasing even in unfluoridated communities, mostly because of fluoride from swallowed toothpaste. A 2009 systematic review indicated that fluorosis is associated with consumption of infant formula or of water added to reconstitute the formula, that the evidence was distorted by publication bias, and that the evidence that the formula's fluoride caused the fluorosis was weak. In the U.S. the decline in tooth decay was accompanied by increased fluorosis in both fluoridated and unfluoridated communities; accordingly, fluoride has been reduced in various ways worldwide in infant formulas, children's toothpaste, water, and fluoride-supplement schedules.

Safety

Fluoridation has little effect on risk of bone fracture (broken bones); it may result in slightly lower fracture risk than either excessively high levels of fluoridation or no fluoridation. There is no clear association between fluoridation and cancer or deaths due to cancer, both for cancer in general and also specifically for bone cancer and osteosarcoma. Other adverse effects lack sufficient evidence to reach a confident conclusion.

Fluoride can occur naturally in water in concentrations well above recommended levels, which can have several long-term adverse effects, including severe dental fluorosis, skeletal fluorosis, and weakened bones; water utilities in the developed world reduce fluoride levels to regulated maximum levels in regions where natural levels are high, and the WHO and other groups work with countries and regions in the developing world with naturally excessive fluoride levels to achieve safe levels. The World Health Organization recommends a guideline maximum fluoride value of 1.5 mg/L as a level at which fluorosis should be minimal.

In rare cases improper implementation of water fluoridation can result in overfluoridation that causes outbreaks of acute fluoride poisoning, with symptoms that include nausea, vomiting, and diarrhea. Three such outbreaks were reported in the U.S. between 1991 and 1998, caused by fluoride concentrations as high as 220 mg/L; in the 1992 Alaska outbreak, 262 people became ill and one person died. In 2010, approximately 60 gallons of fluoride were released into the water supply in Asheboro, North Carolina in 90 minutes—an amount that was intended to be released in a 24-hour period.

Like other common water additives such as chlorine, hydrofluosilicic acid and sodium silicofluoride decrease pH and cause a small increase of corrosivity, but this problem is easily addressed by increasing the pH. Although it has been hypothesized that hydrofluosilicic acid and sodium silicofluoride might increase human lead uptake from water, a 2006 statistical analysis did not support concerns that these chemicals cause higher blood lead concentrations in children. Trace levels of arsenic and lead may be present in fluoride compounds added to water, but no credible evidence exists that their presence is of concern: concentrations are below measurement limits.

The effect of water fluoridation on the natural environment has been investigated, and no adverse effects have been established. Issues studied have included fluoride concentrations in groundwater and downstream rivers; lawns, gardens, and plants; consumption of plants grown in fluoridated water; air emissions; and equipment noise.

Mechanism

Fluoride exerts its major effect by interfering with the demineralization mechanism of tooth decay. Tooth decay is an infectious disease, the key feature of which is an increase within dental plaque of bacteria such as *Streptococcus mutans* and *Lactobacillus*. These produce organic acids when carbohydrates, especially sugar, are eaten. When enough acid is produced to lower the pH below 5.5, the acid dissolves carbonated hydroxyapatite, the main component of tooth enamel, in a process known as *demineralization*. After the sugar is gone, some of the mineral loss can be recovered—or *remineralized*—from ions dissolved in the saliva. Cavities result when the rate of demineralization exceeds the rate of remineralization, typically in a process that requires many months or years.

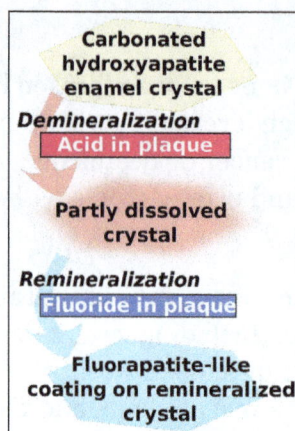

Demineralization and remineralization of dental enamel
in the presence of acid and fluoride in saliva and plaque fluid.

All fluoridation methods, including water fluoridation, create low levels of fluoride ions in saliva and plaque fluid, thus exerting a topical or surface effect. A person living in an area with fluoridated water may experience rises of fluoride concentration in saliva to about 0.04 mg/L several times during a day. Technically, this fluoride does not prevent cavities but rather controls the rate at which they develop. When fluoride ions are present in plaque fluid along with dissolved hydroxyapatite, and the pH is higher than 4.5, a fluorapatite-like remineralized veneer is formed over the remaining surface of the enamel; this veneer is much more acid-resistant than the original hydroxyapatite, and is formed more quickly than ordinary remineralized enamel would be. The cavity-prevention effect of fluoride is mostly due to these surface effects, which occur during and after tooth eruption. Although some systemic (whole-body) fluoride returns to the saliva via blood plasma, and to unerupted teeth via plasma or crypt fluid, there is little data to determine what percentages of fluoride's anticavity effect comes from these systemic mechanisms. Also, although fluoride affects the physiology of dental bacteria, its effect on bacterial growth does not seem to be relevant to cavity prevention.

Fluoride's effects depend on the total daily intake of fluoride from all sources. About 70–90% of ingested fluoride is absorbed into the blood, where it distributes throughout the body. In infants 80–90% of absorbed fluoride is retained, with the rest excreted, mostly via urine; in adults about 60% is retained. About 99% of retained fluoride is stored in bone, teeth, and other calcium-rich areas, where excess quantities can cause fluorosis. Drinking water is typically the largest source of fluoride. In many industrialized countries swallowed toothpaste is the main source of fluoride exposure in unfluoridated communities. Other sources include dental products other than toothpaste; air pollution from fluoride-containing coal or from phosphate fertilizers; trona, used to tenderize meat in Tanzania; and tea leaves, particularly the tea bricks favored in parts of China. High fluoride levels have been found in other foods, including barley, cassava, corn, rice, taro, yams, and fish protein concentrate. The U.S. Institute of Medicine has established Dietary Reference Intakes for fluoride: Adequate Intake values range from 0.01 mg/day for infants aged 6 months or less, to 4 mg/day for men aged 19 years and up; and the Tolerable Upper Intake Level is 0.10 mg/kg/day for infants and children through age 8 years, and 10 mg/day thereafter. A rough estimate is that an adult in a temperate climate consumes 0.6 mg/day of fluoride without fluoridation, and 2 mg/day with fluoridation. However, these values differ greatly among the world's regions: for example, in Sichuan, China the average daily fluoride intake is only 0.1 mg/day in drinking water but 8.9 mg/

day in food and 0.7 mg/day directly from the air due to the use of high-fluoride soft coal for cooking and drying foodstuffs indoors.

Alternatives

The views on the most effective method for community prevention of tooth decay are mixed. The Australian government review states that water fluoridation is the most effective means of achieving fluoride exposure that is community-wide. The European Commission review states "No obvious advantage appears in favour of water fluoridation compared with topical prevention". Other fluoride therapies are also effective in preventing tooth decay; they include fluoride toothpaste, mouthwash, gel, and varnish, and fluoridation of salt and milk. Dental sealants are effective as well, with estimates of prevented cavities ranging from 33% to 86%, depending on age of sealant and type of study.

Fluoride toothpaste is effective against cavities. It is widely used, but less so among the poor.

Fluoride toothpaste is the most widely used and rigorously evaluated fluoride treatment. Its introduction is considered the main reason for the decline in tooth decay in industrialized countries, and toothpaste appears to be the single common factor in countries where tooth decay has declined. Toothpaste is the only realistic fluoride strategy in many low-income countries, where lack of infrastructure renders water or salt fluoridation infeasible. It relies on individual and family behavior, and its use is less likely among lower economic classes; in low-income countries it is unaffordable for the poor. Fluoride toothpaste prevents about 25% of cavities in young permanent teeth, and its effectiveness is improved if higher concentrations of fluoride are used, or if the toothbrushing is supervised. Fluoride mouthwash and gel are about as effective as fluoride toothpaste; fluoride varnish prevents about 45% of cavities. By comparison, brushing with a nonfluoride toothpaste has little effect on cavities.

The effectiveness of salt fluoridation is about the same as that of water fluoridation, if most salt for human consumption is fluoridated. Fluoridated salt reaches the consumer in salt at home, in meals at school and at large kitchens, and in bread. For example, Jamaica has just one salt producer, but a complex public water supply; it started fluoridating all salt in 1987, achieving a decline in cavities. Universal salt fluoridation is also practiced in Colombia and the Swiss Canton of Vaud; in Germany fluoridated salt is widely used in households but unfluoridated salt is also available, giving consumers a choice. Concentrations of fluoride in salt range from 90 to 350 mg/kg, with studies suggesting an optimal concentration of around 250 mg/kg.

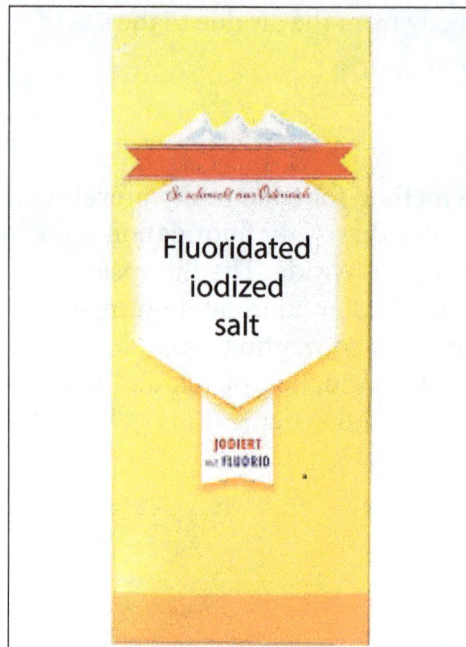

Fluoridated iodized salt sold in Germany.

Milk fluoridation is practiced by the Borrow Foundation in some parts of Bulgaria, Chile, Peru, Russia, Macedonia, Thailand and the UK. Depending on location, the fluoride is added to milk, to powdered milk, or to yogurt. For example, milk powder fluoridation is used in rural Chilean areas where water fluoridation is not technically feasible. These programs are aimed at children, and have neither targeted nor been evaluated for adults. A systematic review found low-quality evidence to support the practice, but also concluded that further studies were needed.

Other public-health strategies to control tooth decay, such as education to change behavior and diet, have lacked impressive results. Although fluoride is the only well-documented agent which controls the rate at which cavities develop, it has been suggested that adding calcium to the water would reduce cavities further. Other agents to prevent tooth decay include antibacterials such as chlorhexidine and sugar substitutes such as xylitol. Xylitol-sweetened chewing gum has been recommended as a supplement to fluoride and other conventional treatments if the gum is not too costly. Two proposed approaches, bacteria replacement therapy (probiotics) and caries vaccine, would share water fluoridation's advantage of requiring only minimal patient compliance, but have not been proven safe and effective. Other experimental approaches include fluoridated sugar, polyphenols, and casein phosphopeptide-amorphous calcium phosphate nanocomplexes.

A 2007 Australian review concluded that water fluoridation is the most effective and socially the most equitable way to expose entire communities to fluoride's cavity-prevention effects. A 2002 U.S. review estimated that sealants decreased cavities by about 60% overall, compared to about 18–50% for fluoride. A 2007 Italian review suggested that water fluoridation may not be needed, particularly in the industrialized countries where cavities have become rare, and concluded that toothpaste and other topical fluoride are the best way to prevent cavities worldwide. A 2004 World Health Organization review stated that water fluoridation, when it is culturally acceptable and technically feasible, has substantial advantages in preventing tooth decay, especially for subgroups at high risk.

Worldwide Prevalence

As of November 2012, a total of about 378 million people worldwide received artificially fluoridated water. The majority of those were in the United States. About 40 million worldwide received water that was naturally fluoridated to recommended levels.

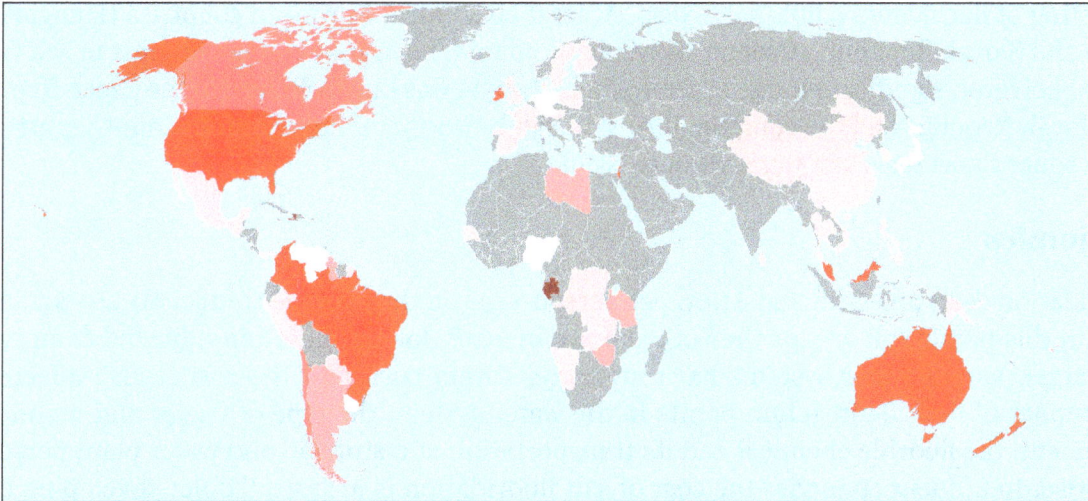

Percentage of population receiving fluoridated water, including both artificial and natural fluoridation, as of 2012. ■ 80–100%, ■ 60–80%, ■ 40–60%, ■ 20–40%, ■ 1–20%, □ < 1%, ▓ Unknown.

Much of the early work on establishing the connection between fluoride and dental health was performed by scientists in the U.S. during the early 20th century, and the U.S. was the first country to implement public water fluoridation on a wide scale. It has been introduced to varying degrees in many countries and territories outside the U.S., including Argentina, Australia, Brazil, Canada, Chile, Colombia, Hong Kong, Ireland, Israel, Korea, Malaysia, New Zealand, the Philippines, Serbia, Singapore, Spain, the UK, and Vietnam. In 2004, an estimated 13.7 million people in western Europe and 194 million in the U.S. received artificially fluoridated water. In 2010, about 66% of the U.S. population was receiving fluoridated water.

Naturally fluoridated water is used by approximately 4% of the world's population, in countries including Argentina, France, Gabon, Libya, Mexico, Senegal, Sri Lanka, Tanzania, the U.S., and Zimbabwe. In some locations, notably parts of Africa, China, and India, natural fluoridation exceeds recommended levels.

Communities have discontinued water fluoridation in some countries, including Finland, Germany, Japan, the Netherlands, and Switzerland. On 26 August 2014, Israel stopped mandating fluoridation, stating "Only some 1% of the water is used for drinking, while 99% of the water is intended for other uses (industry, agriculture, flushing toilets etc.). There is also scientific evidence that fluoride in large amounts can lead to damage to health. When fluoride is supplied via drinking water, there is no control regarding the amount of fluoride actually consumed, which could lead to excessive consumption. Supply of fluoridated water forces those who do not so wish to also consume water with added fluoride. This approach is therefore not accepted in most countries in the world." This change was often motivated by political opposition to water fluoridation, but sometimes the need for water fluoridation was met by alternative strategies. The use of fluoride in its various forms is the foundation of tooth decay prevention throughout Europe; several countries have introduced fluoridated salt, with varying success: in Switzerland and Germany, fluoridated salt represents 65% to 70% of the domestic market, while in France the market share reached 60% in 1993 but dwindled to 14% in 2009; Spain, in 1986 the second West European country to introduce fluoridation of table salt, reported a market share in 2006 of only 10%. In three other West European countries, Greece, Austria and the Netherlands, the legal framework for production and marketing of fluoridated edible salt exists. At least six Central European countries (Hungary, the Czech and Slovak Republics, Croatia, Slovenia, Romania) have shown some interest in salt fluoridation; however, significant usage of approximately 35% was only achieved in the Czech Republic. The Slovak Republic had the equipment to treat salt by 2005; in the other four countries attempts to introduce fluoridated salt were not successful.

Economics

Fluoridation costs an estimated $1.08 per person-year on the average (range: $0.26–$11.46; all costs in this paragraph are for the U.S. and are in 2018 dollars, inflation-adjusted from earlier estimates). Larger water systems have lower per capita cost, and the cost is also affected by the number of fluoride injection points in the water system, the type of feeder and monitoring equipment, the fluoride chemical and its transportation and storage, and water plant personnel expertise. In affluent countries the cost of salt fluoridation is also negligible; developing countries may find it prohibitively expensive to import the fluoride additive. By comparison, fluoride toothpaste costs an estimated $9–$18 per person-year, with the incremental cost being zero for people who already brush their teeth for other reasons; and dental cleaning and application of fluoride varnish or gel costs an estimated $99 per person-year. Assuming the worst case, with the lowest estimated effectiveness and highest estimated operating costs for small cities, fluoridation costs an estimated $17–$26 per saved tooth-decay surface, which is lower than the estimated $98 to restore the surface and the estimated $165 average discounted lifetime cost of the decayed surface, which includes the cost to maintain the restored tooth surface. It is not known how much is spent in industrial countries to treat dental fluorosis, which is mostly due to fluoride from swallowed toothpaste.

Although a 1989 workshop on cost-effectiveness of cavity prevention concluded that water fluoridation is one of the few public health measures that save more money than they cost, little high-quality research has been done on the cost-effectiveness and solid data are scarce. Dental sealants are cost-effective only when applied to high-risk children and teeth. A 2002 U.S. review estimated that on average, sealing first permanent molars saves costs when they are decaying faster than 0.47 surfaces per person-year whereas water fluoridation saves costs when total decay incidence exceeds 0.06 surfaces per person-year. In the U.S., water fluoridation is more cost-effective than other methods to reduce tooth decay in children, and a 2008 review concluded that water fluoridation is the best tool for combating cavities in many countries, particularly among socially disadvantaged groups.

U.S. data from 1974 to 1992 indicate that when water fluoridation is introduced into a community, there are significant decreases in the number of employees per dental firm and the number of dental firms. The data suggest that some dentists respond to the demand shock by moving to non-fluoridated areas and by retraining as specialists.

pH Correction

pH is a measure of the degree of acidity of the water. It is measured on a scale of 0-14 pH units, with low numbers being acidic, 7 being neutral and higher values being classed as alkaline. Many waters are a few units either side of neutral – the regulatory standard for pH in drinking water is within a range between 6.5 and 9.5. Many water supplies derived from source in peaty upland areas, will be slightly acidic due to dissolved organic acids.

Complex Household System (pH filter is smaller, buff coloured unit) – Aberdeenshire Council.

The pH of the water, in itself, is not usually a significant issue – where it can become a problem is where metal plumbing is installed, such as copper piping or galvanized tanks. The acidic water can gradually dissolve the metals, leading to leaks and significant concentrations of the metals in the water itself, potentially at levels harmful to health. Cases of green hair where people have washed hair in private water supplies containing high concentrations of dissolved copper have been widely reported! More commonly, staining of sanitary ware and fittings can occur. Where properties contain old lead pipes, the health aspect becomes even more relevant, as significant concentrations of lead in water can result from relatively small lengths of lead plumbing.

The subject of water corrosivity is a complex one, and a number of factors are involved such as the relative concentrations of different dissolved minerals. In general, however, lower pH waters will tend to be corrosive to plumbing. If metal plumbing is present and the water has a natural pH much below 7, this should be corrected to reduce corrosivity.

The simplest method of elevating pH is to pass the water through a filter bed of alkaline granular material. Such filters are referred to as pH correction filters or neutralizing filters. The alkaline media is usually calcium carbonate or magnesium oxide, or a combination of both. The materials are processed to render them more suitable for water treatment use, and several proprietary brands of pH correction media are available. These neutralizing media, or a blend, will often be used to fill a sediment filter shell. On most supplies, pH correction will usually take place after other filtration stages but prior to UV disinfection. This is probably correct, but in such circumstances, sediment from the pH filter can cause fouling of the UV lamp. If this is the case, a further small filter may be required prior to the UV system. There should be no metal pipework between the pH filter and the UV system to avoid metal deposition on the lamp. If Chlorine disinfection is used, careful thought needs to be given to the siting of any pH correction as higher pH renders disinfection less effective.

Problems can occur due to the fact that apart from elevating the pH, water hardness is also increased. This can give rise to complaints from users especially if they are accustomed to very soft water, its taste and lathering properties. Scale formation can also affect any ultraviolet disinfection installed downstream, reducing its effectiveness.

Small Household System with pH Correction - Aberdeenshire Council.

If the water is free of particulate matter it is not unusual to pass water upwards through the media. The contact time to achieve pH equilibrium can be prohibitively long but times of a few minutes can often have some beneficial effect. Downflow filters can also be used, but these will Complex Household System (pH filter is smaller, buff coloured unit) – Aberdeenshire Council S usually require a backwash system. Even with careful design the alkaline material can clump together. Sediment, organic material and precipitated metals can also coat the media, making it less effective.

When designing a pH correction system for a small water supply, care must be given to sizing and residence times. If filters are undersized and contact times insufficient, the pH will not be sufficiently elevated, however if the water spends too much time in contact with the media it may become too alkaline, with other quality complications. Different media have differing levels of activity, so care must be taken to achieve the correct blend and on-site pilot trials may be advisable. Particular difficulties can arise where a property is only used intermittently through the year, meaning that at certain times, water can spend long periods in contact with the media. Competent specialist advice should be sought.

As a chemical reaction takes place between the water and the media in the neutralizing filter, the media gradually gets used up over time. This means that monitoring of the filter is required and replenishment of the media needs to take place from time to time. If media is simply topped up periodically there is a risk that insoluble deposits will accumulate in the filter and gradually reduce its performance. Therefore, it is good practice to occasionally take the filter off line, clean it out and completely replenish the media. Following replenishment, users may report a slight cloudiness and change in taste, but this should settle after a short amount of time.

References

- What-is-water-disinfection, disinfection, processes: lenntech.com, Retrieved 31 August 2019

- Edzwald, James K., ed. (2011). Water Quality and Treatment. 6th Edition. New York:McGraw-Hill. ISBN 978-0-07-163011-5

- Sheiham A. Dietary effects on dental diseases. Public Health Nutr. 2001;4(2B):569–91. doi:10.1079/PHN2001142. PMID 11683551

- Technical-manual-treatment-ph-correction, media-29308: dwqr.scot, Retrieved 11 April 2019

- Gunn, S. William A. & Masellis, Michele (2007). Concepts and Practice of Humanitarian Medicine. Springer. p. 87. ISBN 978-0-387-72264-1

- Featherstone JD. Dental caries: a dynamic disease process. Aust Dent J. 2008;53(3):286–91. doi:10.1111/j.1834-7819.2008.00064.x. PMID 18782377

Permissions

Index

www.ingramcontent.com/pod-product-compliance
Lightning Source LLC
Chambersburg PA
CBHW061248190326
41458CB00011B/3613